Environmental Science and Engineering

Environmental Science

Series Editors

Ulrich Förstner, Technical University of Hamburg-Harburg, Hamburg, Germany

Wim H. Rulkens, Department of Environmental Technology, Wageningen,
The Netherlands

Wim Salomons, Institute for Environmental Studies, University of Amsterdam,
Haren, The Netherlands

The protection of our environment is one of the most important challenges facing today's society. At the focus of efforts to solve environmental problems are strategies to determine the actual damage, to manage problems in a viable manner, and to provide technical protection. Similar to the companion subseries Environmental Engineering, Environmental Science reports the newest results of research. The subjects covered include: air pollution; water and soil pollution; renaturation of rivers; lakes and wet areas; biological ecological; and geochemical evaluation of larger regions undergoing rehabilitation; avoidance of environmental damage. The newest research results are presented in concise presentations written in easy to understand language, ready to be put into practice.

More information about this subseries at http://www.springer.com/series/3234

Maryam Pazoki · Reza Ghasemzadeh

Municipal Landfill Leachate Management

Springer

Maryam Pazoki
College of Environmental Engineering
University of Tehran
Tehran, Iran

Reza Ghasemzadeh
College of Environmental Engineering
University of Tehran
Tehran, Iran

ISSN 1863-5520 ISSN 1863-5539 (electronic)
Environmental Science and Engineering
ISSN 1431-6250 ISSN 2661-8222 (electronic)
Environmental Science
ISBN 978-3-030-50214-0 ISBN 978-3-030-50212-6 (eBook)
https://doi.org/10.1007/978-3-030-50212-6

This Springer imprint is published by the registered company Springer Nature Switzerland AG
The registered company address is: Gewerbestrasse 11, 6330 Cham, Switzerland

Contents

List of Figures

List of Tables

Chapter 1
Waste Management

1.1 Introduction

1.1.1 History of Waste Management

Solid wastes left on the land by early human beings probably remained of hunting or other surviving activities. Human excrement accounted for another problem associated with leftover materials during prehistoric times. Ancient nomadic people used to migrate to other places once wastes have piled. Then these wastes were absorbed in nature through innate mechanisms such as scavenging and decaying. Therefore, the dominant environmental issues of modern times like air or water pollution were rather negligible during prehistoric times, since the very low population of human lived on the Earth then.

Around 9000 BCE, the early man decided to put an end to his nomadic life and settle down in one place perpetually. Humans changed their lifestyles from hunting to farming and crafting, so they became more socially-organized. The permanent settlement had some undesirable consequences due to the considerable increase in waste materials and their accumulation for longer times. The new communities were confronted with new problems of increasing accumulation of remains, so they had to adopt a method to deal with these issues (Tchobanoglous et al. 1993a).

The obtained ancient stuff belonging to prehistoric humans such as different types of tools, weapons, and artifacts have provided us with a better knowledge of diverse eras including the Stone Age, the Bronze Age, and the Iron Age.

Archaeologists have acquired a good knowledge about the prehistoric inhabitants of the Earth by digging into the leftover waste-yards and examining the stuff that remained in the old towns. These investigations have provided them with useful information about social, cultural, and traditional codes, and also about the eating habits of the ancient communities. For example, they have discovered non-degradable objects like instruments, weapons, and appliances that have perceived as belongings

© Springer Nature Switzerland AG 2020
M. Pazoki and R. Ghasemzadeh, *Municipal Landfill Leachate Management*,
Environmental Science and Engineering,
https://doi.org/10.1007/978-3-030-50212-6_1

to Stone Age people. Mayans from Central America used to place their impaired decorative accessories, tools, and home appliances they no longer needed into their regal graves whenever they faced stagnant economic situations (Tchobanoglous et al. 1993a). However some disposed of materials were reused for other applications, for example, some broken pieces of earthen containers and utensils have been found in building blocks of some shrines (Pazoki et al. 2016).

1.1.2 Waste Management System

Typical operations involved in any solid waste management system are waste reduction (causes, amount, formation, and quality), collection, handling and transportation, treating, and finally disposal. Waste reduction, reuse, and recycling are typically integrated and combined in comprehensive waste management procedures. The components of the process and coordination among them should be taken into account individually during delineating the waste management system (Ghasemzade and Pazoki 2017).

For example, Indonesian policymakers have reckoned a one-third reduction in the amount of waste reuse through enforcing more constraints on rubbish scavengers.

As the developing countries are becoming more industrial and modern, components of wastes change accordingly, for instance, banana leaves have been substituted with plastic bags throughout East Asian countries.

People from developing countries are not so keen to dispose of their household garbage separately since they don't find the task suited to their social position. Non-separated wastes result in some serious problems in recycling operations due to inconsistencies in the compositions of the materials, and also because the materials spoil and degenerate easily under such conditions (Tchobanoglous et al. 1993a; Ludwig et al. 2012).

In contrast to procedures in which wastes are extracted from mixed rubbish mass, recycling operations by adopting the approach of separating garbage according to materials origins yield much more materials of more favorable quality.

Other relevant methods in this field involve waste reduction, reuse, recycling, and recovery (the 4Rs) which has been designed to lower the disposal costs, and enforcing more restraints on environmental effects due to waste dumps. 4Rs has some advantages over other methods because of being able to bring about decreases in greenhouse gases; air, land, and water pollution; the quantity of to-be-disposed waste on one side and saving water, energy, and resources on the other side (Ludwig et al. 2012).

Nevertheless, there are lots of solid wastes that have not been brought into this sequence yet, and it implies the fact that there is still a large number of wastes to be recycled.

Some objects are being sold to scavengers who vend them in turn to mass recycling plants and producers.

1.2 An Outline of Municipal Solid Waste Management (MSWM)

In this section, an overall overview of the urban solid waste treatment process is presented providing a comprehensive description of the involved elements and dynamics in this process. A summary of the common concepts including waste categorizations and functional elements of MSWM is also introduced here. The major target is providing readers with clarification on MSDM, its functions, and coordination among them.

1.2.1 Descriptions

1.2.1.1 Solid Waste

Solid wastes are undesirable and useless for producer results of human and animal activities including industrial, social, and agricultural activioties. Solid waste is an inevitable outcome of living that has various forms within different communities (Letcher and Vallero 2019; Delarestaghi et al. 2018).

Municipal solid waste refers to the portion of solid waste which is produced in urban areas. This kind of waste is mostly composed of leftovers, wrapping and packaging items, bottles and containers made up of glass, paper, cans, and PET (Tchobanoglous et al. 1993a, b). The increasing amount of solid waste in most developed societies following the Industrial Revolution is a consequence of major alterations made to using habits of such communities (Tchobanoglous et al. 1993a, b).

Municipal solid waste is mostly characterized by packaging items that are typically used for a wide range of goods consumed for modern life. Plastic and cardboard play essential roles in packaging, so they make a considerable portion of waste in our routine life (Vesilind et al. 2002).

1.2.2 Solid Waste Management

Any typical management system for solid wastes can be involve waste resuction, source minimization, managing in materials that used for generation, on-the-field handling and sorting, storage and arranging, collection and transpotation, recycling, recovery, disposal, and post-disposal monitoring (Pazoki et al. 2015a).

Solid waste management can refer to all efforts associated with solid waste. They may include retention, accumulation, conveyance, treating, processing, and final disposal.

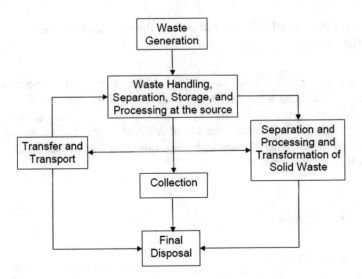

Fig. 1.1 Solid waste management functions and their interrelationship

Some significant issues including technological effects, economic and social factors, and public health need to be taken into account while defining management plans. The presence of various factors from multiple fields adds further complications to the task of waste management (Tchobanoglous et al. 1993a, b) (Fig. 1.1).

Solid waste management involves the following functional elements or tasks: (1) waste generation, (2) waste handling, separation, storage, and on-site treatment, (3) collection, (4) separation, treatment, and recycling of the waste, (5) transfer and transport, and (6) disposal (Pazoki et al. 2015b).

1.2.2.1   Waste Production

All pertinent actions taken for using needed goods and then getting rid of worthless and undesirable materials are considered as waste production (Fig. 1.2).

It is noteworthy that managing the waste production stage can be highly challenging. It requires a lot of effort for creating public knowledge, informing, teaching, guiding, approval, enforcing communal penalizing, and so on (Tchobanoglous et al. 1993a).

1.2.2.2   Waste Handling, Separation, Storage, and On-Site Treatment

This practice, by itself, encompasses controlling and separating tasks engaged in transferring and keeping the waste within specially-made containers. handling involves transportation of the full containers to the accumulation area. A critical

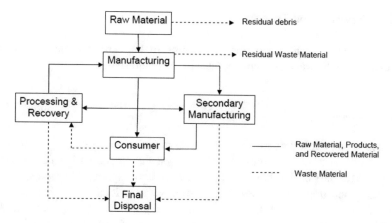

Fig. 1.2 Materials flow and waste production in society

task here is initial separation at the origin. It can not only make the waste ready for recovery and recycling but also minimizes the risks of waste. MSWM process often begins here in many modern communities (Parsa et al. 2018; Ghasemzadeh et al. 2017).

1.2.2.3   Collection

The operation entails gathering solid waste and the separated substances, then carrying them to the already prepared areas in which collection machines should be unloaded. The task is often highly capital-intensive. The length of transportation trips to the unloading sites and quantity of waste are two major factors affecting the process (Kalamdhad et al. 2018).

1.2.2.4   Separation, Treatment, and Transformation of Solid Wastes

Separation and treatment are general terms given to operations that can be divided further into some sub-stages: recovering, refining, burning of the detached waste, ripping manually or using special equipment, detaching iron-contained materials by making use of magnets, reducing the size of waste by squeezing and burning (Ghosh 2018).

Recycling and Recovery refers to a physical, chemical and biological process through which the under-treatment waste is diminished and converted to a reusable substance. A recognizable example of such a task is generating biogas in the course of anaerobic decomposition (Shayesteh et al. 2020; Pazoki et al. 2018).

1.2.2.5 Transfer and Transport

This stage involves transferring the waste from smaller carriers to large trucks and also conveying the waste from piling sites to treatment stages.

1.2.2.6 Disposal

This step usually accounts for the final phase of MSWM. A certain portion of the useless waste is taken to dumps directly, and yet the remaining batches are transferred to the same place following recycling and extracting reusable products. Environmental impacts and public health considerations are critical issues that need to be taken into account in this field. However dumps are not only critical points regarding public health, but they also deserve meticulous attention for production requirements in the future (Reddy et al. 2011; Pazoki et al. 2017).

1.2.3 Hierarchy of Waste Management

Any waste management system consists of certain identified plans that have to be prioritized according to importance, thus executive levels are highly significant. As depicted by Fig. 1.3 below, the waste management system encompasses the elimination of source, recovering of reusable materials, waste transformation, and finally disposal (Tchobanoglous et al. 1993a; Pazoki et al. 2015b).

Levels of waste management can be defined from various perspectives. However, a holistic view considers 4 sub-divisions which incorporate materials reusing as a component of recycling. Yet other perspectives suppose elimination of source, retreatment, energy recovery, and final disposal as the fundamental levels for waste

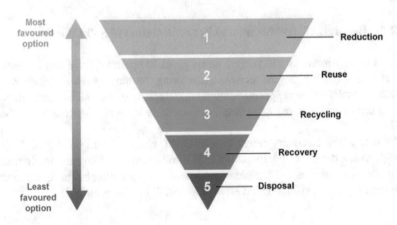

Fig. 1.3 Hierarchy of waste management

management. Nevertheless, what really matters is to develop a definite policy and establish continuous management throughout the community (Tchobanoglous et al. 1993b).

1.2.3.1   Source Reduction

It accounts for an essential part of the waste management system. It refers to preventing waste production on a large scale. Minimization requires both a decrease in the quantity of the waste and restricting allowable ranges for its harmfulness. This objective may be realized during production by making modifications in design and packaging, and it is also obtainable by households by modifying their consumption patterns (Cheremisinoff 2003).

In order to actualize a fruitful source reduction the following items are to be taken into consideration (Tchobanoglous et al. 1993b):

1. Recounting the waste produced in each phase and deciding on the relevant source reduction potential
2. Imposing a levy on disposal products
3. Supplying tax credits or exemptions to industries that meet set source reduction goals in design and production
4. Set the source reduction database in each part (household, industrial, commercial)
5. Amend legislative rules toward proposing more incentives for waste reduction by diminishing costs for submissive businesses.
6. Enforce corrective measures about components, commodities, and entities that have brought about the widest environmental and economic consequences.

1.2.3.2   Recover Reusable Materials—Recycling

The second most critical level of the waste management system belongs to recycling. It is commonly perceived as an effective utility capable of minimizing the exhaustion of resources and reducing adverse effects by decomposing wastes and turning reusable materials back to the consumption cycle. In technical terms, recycling refers to all operations required for separation and collection of reusable substances in order to return them to the cycle of producing and consuming (Worrell and Reuter 2014).

1.2.3.3   Transformation

The next level following recycling in significance ratings belongs to transformation. Once the reusable materials have been recovered, they have to be reprocessed and converted to more usable compositions through chemical, physical, and biological processes or even burning.

1.2.3.4   Landfilling

The final stage of waste management is getting rid of materials which are not possible to be recovered or transformed. What goes to landfills is unrecyclable material remained from preceding processes.

1.2.4   Varieties and Characteristics of Urban Solid Waste

1.2.4.1   Types and Sources

In the midst of managing MSW, there are some fundamental subjects that should be considered, such as:

Source, kind, and configuration of waste in each section; quantity and rate of waste generation, dangerous constituents in waste and so forth.

We can identify the origins of solid waste under 8 classifications. They include business-driven, urban services, organizational, building and destruction, industrial, agricultural, residential, treatment plants. Table 1.1 shows a summary of the types and resources of the MSW (Tchobanoglous et al. 1993a, b).

1.2.4.2   Characteristics

Recognizing the physical, chemical and biological potentialities of wastes is vital for better setting about and treating the waste. The adopted methods to accumulate, store, separate, transport, and carry the waste depend on physical properties of wastes such as specific weight, granules size, distribution of size, the porosity of pressed waste, and content of humidity. Knowledge about the chemical composition of materials can be influential in determining the required chemical processes. For example, the amount of emancipated energy in the result of combustion may be estimated on the basis of chemical characteristics of the waste. Wastes mostly contain carbon, hydrogen, oxygen, nitrogen, sulfur, ash, and some other elements with low contents.

Wastes have biological potentials as well. Organic contents of solid wastes are used as raw materials for obtaining biogas or making fertilizers. Similarly, acquiring awareness about nutrients and valuable components that exist in the process is equally significant. Smell generation and quality of being prone to biological decomposition are two main characteristic factors in this context.

1.2.4.3   Hazardous Waste

Any waste that may harm public or environmental health is regarded as hazardous waste. Identification and appropriate treatment of such wastes are two critical issues in MSWM, since they may have continuous adverse impacts for a considerably long

Table 1.1 Source of solid wastes within a community

Source	Typical facilities, activities, or locations where wastes are generated	Types of solid wastes
Residential	- Single-family and multifamily detached dwellings - Low-, medium-, and high-rise apartments, etc.	Food wastes, paper, cardboard, plastics, textile, leather, yard wastes, wood, glass, tin cans, aluminum, other metals, ashes, steel leaves, special wastes (incladding bulky items, consumer electronics, white goods, yard wastes collected separately batteries, oil, and tires), household hazardous wastes
Commercial	Stores, restaurants, markets, office buildings, hotels, motels, print shops, service stations, auto repair shops, etc.	Paper, cardboard, plastics, wood, food waste, glass, metals, special wastes (see above), hazardous wastes, etc.
Institutional	Schools, Hospitals, prisons, governmental centers	As above in commercial
Construction and demolition	New construction sites, road repair/renovation sites, razing of buildings, broken pavement	Wood, steel, concrete, dirt, etc.
Municipality services (excluding treatment facilities)	Street cleaning, landscaping, catch basin cleaning, parks, and beaches, other recreational areas	Special wastes, rubbish, street sweepings, landscape, and tree trimmings, catch basin debris, general wastes from parks, beaches, and recreational areas
Treatment plant sites; municipal incinerators	Water, wastewater, and industrial treatment processes, etc.	Treatment plant wastes, principally composed of residual sludge
Municipal solid waste[a]	All of the above	All of the above
Industrial	Construction, fabrication, light and heavy manufacturing, refineries, chemical plants, power plants, demolition, etc.	Industrial process wastes, scrap material, etc. Non-industrial wastes including food wastes, rubbish, ashes, demolition and construction wastes, special wastes, hazardous wastes
Agricultural	Field and row crops, orchards, vineyards, dairies, feedlots, farms, etc.	Spoiled food wastes, agricultural wastes, rubbish, hazardous wastes

[a]Generally, the term urban solid waste includes all waste produced in society except for wastes generated through industrial and agricultural processes

period. Perilous contents within wastes can threat environmental and public health almost in all stages from collecting through processing to final dumping.

Hazardous wastes have various properties that can be classified into flammable, corroding, reactive, burnable, poisonous, and toxic. Considering the adverse effects they may have brought about, the process of treating hazardous wastes need to be in isolated conditions apart from other types of waste (Sengupta and Agrahari 2017).

Hazardous wastes commonly abound in business-bound and residential areas and are so frequent in cleaning agents, cosmetics, and industrial products such as automotive oils, solvents, and paints. Residents of some developed countries are now so vigilant and aware of potential risks that these products may have, they have been accustomed to separate them from other wastes and discard hazardous litters to specially-designed dumps (Martin et al. 2013).

1.3 Elements of MSWM

As previously mentioned, there are several operations involved in solid waste management that are highly complicated. A summary of these operations is introduced here to illustrate individual processes.

1.3.1 Waste Production

Designing the collecting routes, recycling and disposal utilities depends on the amount of solid waste that has to be estimated. We can apply various approaches to quantify waste amounts. The common methods for the purposes of this task may include load count, weight-volume, and material analyses. The load count and material analyses are both based on estimating the quantity of waste during a specific time. In the case of material analysis, a much larger set of data is required to be examined. Some common data for this purpose are the quantity of waste to be recovered, contents of ash and conduit gas within combustion chambers. The present method requires a boundary to be previously defined, then all input and outputs will be examined. By doing in this way, the quantity of waste will be measurable.

In order to gain a total estimation of the waste quantity, all collected data should be statistically analyzed. As discussed before, the urban waste may originate from various sources including business, residential, construction, and destruction, industrial, agriculture, refinery plants, etc.

Several elements may take part in waste production. Elimination of source accounts for a significant factor. Elimination of source refers to all corrective actions taken during designing, producing, packaging, and distributing stages aiming at reducing the amount of waste. Restricting the packaging industry to the minimum consumption of material or developing products with a longer life are two critical issues contributing to this objective. Reprocessing is another element playing a critical role in waste management. In addition to the mentioned technical factors, there are some subsidiary aspects associated with public viewpoints and national regulations.

1.3.2 Waste Handling, Seperation, Storage, and Processing at the Source

Waste separation accounts for a significant process in the overall waste management system. Waste treating refers to all handling operations executed before being transferred to the containers from which wastes are delivered to subsequent processes for recycling. These operations facilitate being-processed waste flow throughout the system. The separating operations in order to pick up papers, cardboards, aluminum containers, plastics, etc. play an important role in minimizing the costs of required arrangements and equipment in the recycling stations. Various approaches are available for separate operations. In the case of producing points within residential areas, the approach to management may vary depending on the involved factors including types of the buildings, local population, and methods applied by the population for collecting waste. Residents of low-dense regions tend to gather their garbage inside the home, then they put them in special containers placed in the neighborhood; however, in densely populated high apartments, rubbish chutes have been installed to pile wastes from resident households. In the case of trading and industrial estates, special containers are placed near the waste-producing points depending on the type and volume of waste. Then the accumulated garbage is loaded into larger containers that are carried to disposal stations. It should be noted that the type and capacity of containers are important factors that need to be selected reasonably for appropriate transportation. Moreover, there are other critical issues that have to be taken into account: sources from which wastes are produced, the quantity of waste, intervals between transportations, and problems due to the generated smell. Under certain conditions, applicable resolutions are provided which allow recycling operations at the origin. For example, some composting techniques have been invented that enable producing composts from potential parts of the waste directly. Supplying combusting devices to burn the waste at source is another solution. Since the cost of collecting and transferring operations can be higher in trading and industrial estates, tearing and compacting equipment may also be provided to perform preparatory tasks prior to collecting (Tchobanoglous et al. 1993a, b; Cheremisinoff 2003).

1.3.3 Solid Waste-Collecting

Collecting operations are among the most difficult missions in solid waste management. The job involves both collecting sold wastes from producing points and carrying them to successive processing sites. Collecting operation even becomes more complicated if there is no sorting system provided at the origin. This operation accounts for the most costly stage of the waste management system. Handling operations at the source are highly significant since they determine the collecting operations. The collection system design depends on the type of waste—separated or un-separated—the distance between collecting and disposal sites, and paths leading to the sites (Delarestaghi et al. 2018).

1.3.3.1 Collecting Mixed Wastes

The preventive measures aiming at reducing such wastes from residential areas may be accomplished in four ways (Tchobanoglous et al. 1993a):

1. Pavement: one special day for collection is defined when residents in the neighborhood are supposed to put their full garbage containers in predefined locations where trucks will collect them. Then they return the container back once they have been unloaded.
2. Pathway: locations determined on some specified parts of the city that containers for garbage storage are located.
3. Set out—set back: residents are not responsible for transferring full and empty containers; they only need to make them ready for waste-carrying staff who will return the unloaded containers back to the first place as well.
4. Set out: the procedure is similar except that residents themselves are responsible for empty containers after being unloaded.

Similarly, full containers may be emptied into carrying trucks through various methods.

Full containers are directly picked up by hands or with the aid of special mechanical devices and lifts in order to be unloaded into carrying trucks and transported to processing sites.

However, in the case of multistory apartment buildings specially-designed mechanical collecting systems are provided depending on the size and design of garbage containers.

Collecting operations in trading and industrial estates are carried out by special trucks that are capable of picking up full containers and replacing empty ones to be loaded for next time.

1.3.3.2 Collecting Seperated Wastes from Producing Sources

Collecting operations in the residential areas depend on methods of separating wastes in society. For example, two types of containers are provided for recyclable litters such as papers and plastics, and one container for other materials. In some other societies, more containers are provided for separating various types of recyclable wastes. Normally, distinct tracks have been assigned to each individual container to be transported to processing sites. A person is usually contracted to handle recyclable wastes (Ludwig et al. 2012).

1.3.3.3   Various Ways of Collection

Waste collection operations are usually accomplished by a variety of containers and carrying systems. Means of transportation used for this purpose are generally of two main types: Freight container and stationary container (Tchobanoglous et al. 1993a).

(a)   **Freight-container system**

Such a system is more appropriate for sites where large quantities of waste are processed. Doing in this way has advantages of reducing the collections rate and collection costs accordingly. Moreover, they are available in diverse sizes and shapes catering to various types of waste. These systems are usually provided with the tow truck, tilt-frame containers, and especially-designed garbage systems.

Systems that are equipped with tow-trucks are applicable for small operations with only a few pick-up points. Such systems are particularly proper for compressible metal pieces and building materials.

Tilt-frame systems are suitable for larger containers handling enormous quantities of the waste. Some systems are equipped with compacting devices that diminish the volume of the waste and allow carrying much more garbage in a single operation. Thus, waste management by such utilities will be more productive.

The basic structure of especially-designed garbage systems is similar to that of tilt-frames, but it's mainly designed for heavy waste like sand, metal scraps, and also wastes from construction or demolition operations.

Normally, one driver is responsible for handling and carrying operations. However, some situations particularly handling hazardous wastes necessitate the presence of an auxiliary worker who helps the driver, so the number of required workers is typically two at last.

(b)   **Stationary-container system**

This system can be used for all modes of solid waste management. It may be applied manually or mechanically operated containers. Sometimes for the purpose of minimizing the costs of transportation operations, once the wastes have been collected from several sources, they are loaded into a larger container to be carried to the processing sites. Since plenty of time is available for loading of containers, manual operation is typically more effective than the mechanical one. This system requires one driver and an auxiliary collector; however, two collectors work together depending on the area size.

It should be noted that analyzing the collection system and designing commuting paths are highly significant for planning the waste collection system in terms of management. As the commuting path gets longer in both time duration and distance, the costs increase accordingly. Path design is an issue that is beyond this book's scope.

1.3.4   Segregation, Processing, and Transformation of Solid Waste

The fourth stage within the structure of solid waste management deals with segregation, processing, and transformation of solid wastes.

This function is accomplished in material recycling facilities (MRF) or material recycling/transfer facilities (MR/TF) which engage integrated stations for separating wastes, material separating units, biological treatment units, and transforming units to process wastes into fuel.

This stage of MSWM aims at separating recyclable and reusable materials from the pile based on the classifications that types of waste fit into. This objective is realized in 5 stages (Tchobanoglous et al. 1993a, b):

1.   Direct separation and recovering of materials such as wooden pallets, lumber, gal drums and etc.
2.   Using the obtained materials like papers, glass, cardboard, cans and etc. as raw materials for manufacturing and reprocessing.
3.   Using materials with transformational potentials such as leftovers, garden wastes and etc. as feedstocks for biological composting.
4.   Using reusable wastes as bases for producing fuels by converting them to biogas and bio-ethanol, and also extracting heating energy from the wastes by burning them.
5.   Transporting the residues, as minimum as possible, to the landfills.

1.3.4.1   Waste Segregation

As already mentioned, the most desirable approach to waste segregation is separation at the source. Two common methods can meet this need: Drop-off sites and Buy-back sites. Drop-off sites refer to special places with the availability of separation utilities that provide residents with easy access to processing stations; they only need to put their garbage into specially-designed containers. Drop-off stations are provided with distinct containers for each type of waste including recyclable, reusable, compostable, and etc. Some problems arise from densely populated regions with a large number of wastes, thus Drop-off stations are usually set up adjacent to business centers or residential areas not only to diminish transportation costs but also to encourage more participants to take part. Buy-back stations are similar to drop-off sites in terms of operational principles, but here participants have to pay in proportion to the number of wastes they have produced. Such a policy may courage cutting the way down on producing wastes.

Furthermore, there are some other applicable methods for separating the wastes:

1. Every resident can separate recyclable waste manually at the source.
2. Separations that carried out at MRF or MR/TF.

Any recyclable and reusable material will be separated from the pile of mixed wastes. These utilities are also responsible for ensuring the desired quality of the recovered materials.

Production in Material Recycling Facilities (MRFs)
As explained in the earlier section, MRFs are places where required management processes are carried out. All measures and equipment supplied by these facilities depend on the defined objectives of waste treatments. MRF missions depend on: (1) contribution they may make to waste management, (2) types of the recovered materials, (3) delivery conditions of both waste and recyclable materials, and (4) storage of the recovered materials (Lund et al. 1994).

Similarly, the design of MRFs certainly varies from separated materials to mixed wastes. Certain engineering evaluations are required prior to designing. Once the objective of MRF has been determined, the next phase will be the identification of materials to be separated meeting the objective. The next stage involves proceeding with the process according to a predefined flow diagram. This flow diagram should indicate the flow of materials from the moment they are delivered to processing stations until they are discharged. The capacity of the facilities in terms of physical, chemical, and biological transformations of the wastes is chosen based on the process rating load which has been defined in advance. The facilities are to be tailored to the objectives of MRF in the waste refinery. Moreover, environmental and aesthetic aspects are critical issues that have to be taken into account in MRFs design. For instance, surface or ground waters, climatic variations, public health, problems with an unpleasant smell, and emissions are some of the critical considerations. The last issue that concerns us is whether the adopted measures can meet future requirements by modifying the production rate and improving other conditions.

1.3.4.2 Waste Recycling

Recycling refers to operations aimed at minimizing the amount of waste disposal by converting as much as possible recyclable materials to usable products or extracting maximum energy from them. These functions may be accomplished by biological and chemical processes.

1.3.5 Transportation

Transportation includes all tasks dealing with conveying wastes from one place to another. Wastes are not usually transferred directly from sources to MRFs since intermediary facilities such as loaders, trucks, and conveyors cooperate as well.

These facilities are used two reasons: firstly, MRFs are far from waste sources, and they also improve outputs from waste materials. Generally, the facilities are designed to meet these elements in MSWM: (1) avoiding illicit dumping because of long carrying distances, (2) disposal sites are usually away from collection paths, (3) making maximum use of capabilities provided by equipment, (4) shortage of services in residential areas, (5) using the hauled container system, (6) to take advantage of hydraulic and pneumatic mechanisms in collection operations.

1.3.6 Disposal

The process of MSW ends in the disposal phase which involves dumping unusable wastes—those that cannot be recycled or be subjected to energy-acquiring biological or chemical process—into landfills. Besides, central governments of many developed or developing countries impose taxes on waste dumping, so the issue of waste disposal accounts for critical concerns in both environmental and economic considerations.

1.3.6.1 Design of Landfills

Designing landfills is a complicated procedure due to the critical environmental impacts that they may cause. Protecting the environment against the destructive effects of landfills requires some factors that are summarized below (Bagchi 2004):

(a) **Design and layout of landfills**

Procedures and types of operations in landfills depend on the quantity and type of wastes that are to be put in these places. The type of operations depends on the type of wastes and local conditions. Various landfills for hazardous wastes, for restricting biogases production, and for the wetland are some typical examples that necessitate unique elements to be taken into consideration for designing. The location of landfills, the distance between sources and landfills, properties of soil and climatic conditions, area of land, specifications of surface and ground waters, and local regulations are some of the other factors that can play significant roles in the design of landfills.

(b) **Operation and management in landfills**

Since various kinds of waste materials may be found in landfills, different interventions may be required in dealing with them. Biological and chemical processes bring about certain problems related to undesirable environmental effects such as harmful emissions, contaminating leaching products, and unpleasant smells. The major issue in the field of landfill operations is restraining these influences. As an instance, treating the emitted gas from landfills to produce methane is an effective solution to make use of this energy. Gathering by-products from leaching and treating them to minimize their harmful effects before releasing to the environment is another example of the preventive measures. Enforcing controls over the surface water is another approach that is usually done by applying covers for each layer of landfills.

Furthermore, some special equipment is also necessary for the landfills to accomplish the jobs of transportation, excavation, compacting, and combining. High-track compactors with rubber tire and a loader on the front end are common in landfills.

(c) Environmental monitoring

The environmental impacts of landfills are required to be monitored continuously. Groundwater, emitted gases, and leaching products are significant elements that need to be considered in monitoring. Some monitoring measures are also necessary during cessation operation and after it. Landfills are usually monitored for 30 to 50 years after closure.

(d) Safety

Safety is a general term that encompasses public health, occupational health of the workers in terms of safety measures adopted in the working places, landfills here. Considering the vast area covered by landfill operations, the presence of enormous vehicles, and the health impacts of the landfill itself, developing definite regulations regarding workers, working conditions, and workplace situations is necessary.

1.4 Solid Waste Management in Developing Countries

Municipal solid waste management brings about serious problems in many cities in developing countries. Rapid augmentation in population has resulted in a key soar in waste generation. The economic recession in many large cities causes some challenges to effectively managing solid waste. Concept of Integrated Solid Waste Management was created in the developed countries. Executing effective solid waste management in the developing world requires modifications (Durand 2013).

Practices for waste management—collection, processing, and disposal—vary from country to country, generally according to the features of the waste stream and according to local environmental and economic characteristics. Developing countries hold certain socio-economic situations in common. These involve rapid population growth, migration to urban regions, lack of sufficient funds and affordable services, and low-skilled labor forces. These countries are faced with serious environmental and administrative challenges with respect to solid waste management. In developing countries, one can find unworkable waste management trends, which end in negative influences on environmental quality and public health. In general, organization and planning of waste collection systems in developing countries is very rudimentary; solid waste management systems are poorly run and operate at low standards due to insufficient and unfeasible programming. Lots of cities do not have any data on the amount and type of waste produced; neither do they have the resources to conduct such studies. Also, many societies still lack proper disposal sites for the waste produced and do not have any recovery and or recycling programs (Durand 2013; Zurbrugg 2003).

Strengthened institutions and/or new institutional patterns are required for effective waste management services in Third World cities. The purpose of this procedure is to assemble and dispose of solid waste generated in an environmental and socially satisfying manner, with the least probable use of economic resources.

Third world cities spend up to 40% of their budgets trying to improve their waste management programs. However, policymakers stay in line with other problems like air pollution or wastewater treatment than to solid waste management. Often programs that do get executed include solutions that are centralized and bureaucratic, with little public association in the decision process.

Another challenge that municipalities, in developing countries confront is the lack of administrative organizations between departments. Typically, Municipal Governors are responsible for many urban services at the same time. Excessive accountability has a negative impact both on the quality of the services presented and on the allotment of the financial resources within the different services. As a result, many of these services end up being allocated to the staff with low education levels and no technical training in the specific field. Likewise, due to low incomes paid in this sector, many employees do not seek to achieve the best results possible in the functions assigned to them. This lack of training is also reflected in the generation of overly ambitious public by-laws that miss proper monitoring and control procedures. This creates a huge gap between policy and practice.

Generally, in developing countries, solid waste management is scarce and insufficiently planned. There is an abundance of unskilled labor, thus labor is cheap. This makes for a high rate of labor-intensive collection and managing of recyclable things throughout the developing world. A common procedure in developing countries' waste management systems is that collection workers often sift through household waste handed in for compilation; there are also scavengers who sift through waste at transmitting stations and dumpsites.

These suggests that there is a need to modernize solid waste management by implementing recycling or reuse programs. Also, solid waste management plans should implement treatment systems that aim to counteract environmental vicissitudes, cutting the consumption of natural resources, and minimizing the space required for final disposal.

References

Bagchi A (2004) Design of landfills and integrated solid waste management. Wiley
Cheremisinoff NP (2003) Handbook of solid waste management and waste minimization technologies. Butterworth-Heinemann
Delarestaghi RM, Ghasemzadeh R, Mirani M, Yaghoubzadeh P (2018) The comparison between different waste management methods of Tabas city with life cycle assessment assessment. J Environ Sci Stud 3(3):782–793
Durand C (2013) Sustainable waste management challenges in developing countries. Paris, France

Ghasemzade R, Pazoki M (2017) Estimation and modeling of gas emissions in municipal landfill (Case study: Landfill of Jiroft City). Pollution 3(4):689–700

Ghasemzadeh R, Pazoki M, Hoveidi H, Heydari R (2017) Effect of temperature on hydrothermal gasification of paper mill waste, case study: the paper mill in North of Iran. J Environ Stud 43(1):59–71

Ghosh SK (ed) (2018) Waste management and resource efficiency: proceedings of 6th IconSWM 2016. Springer

Kalamdhad AS, Singh J, Dhamodharan K (eds) (2018) Advances in waste management: select proceedings of recycle 2016. Springer

Letcher TM, Vallero DA (eds) (2019) Waste: a handbook for management. Academic Press

Ludwig C, Hellweg S, Stucki S (eds) (2012). Municipal solid waste management: strategies and technologies for sustainable solutions. Springer Science & Business Media

Lund JR, Tchobanoglous G, Anex RP, Lawver RA (1994) Linear programming for analysis of material recovery facilities. J Environ Eng 120(5):1082–1094

Martin WF, Lippitt JM, Prothero TG (2013) Hazardous waste handbook for health and safety. Butterworth-Heinemann

Parsa M, Jalilzadeh H, Pazoki M, Ghasemzadeh R, Abduli M (2018) Hydrothermal liquefaction of Gracilaria gracilis and Cladophora glomerata macro-algae for biocrude production. Biores Technol 250:26–34

Pazoki M, Pari MA, Dalaei P, Ghasemzadeh R (2015a) Environmental impact assessment of a water transfer project. Jundishapur J Health Sci 7(3)

Pazoki M, Delarestaghi RM, Rezvanian MR, Ghasemzade R, Dalaei P (2015b) Gas production potential in the landfill of Tehran by landfill methane outreach program. Jundishapur J Health Sci 7(4)

Pazoki M, Abdoli MA, Ghasemzade R, Dalaei P, Ahmadi Pari M (2016) Comparative evaluation of poly urethane and poly vinyl chloride in lining concrete sewer pipes for preventing biological corrosion. Int J Environ Res 10(2):305–312

Pazoki M, Ghasemzade R, Ziaee P (2017) Simulation of municipal landfill leachate movement in soil by HYDRUS-1D model. Adv Environ Technol 3(3):177–184

Pazoki M, Ghasemzadeh R, Yavari M, Abdoli M (2018) Analysis of photocatalyst degradation of erythromycin with titanium dioxide nanoparticle modified by silver. Nashrieh Shimi va Mohandesi Shimi Iran 37(1):63–72

Reddy KR, Hettiarachchi H, Gangathulasi J, Bogner JE (2011) Geotechnical properties of municipal solid waste at different phases of biodegradation. Waste Manage 31(11):2275–2286

Sengupta D, Agrahari S (eds) (2017) Modelling trends in solid and hazardous waste management. Springer, Singapore

Shayesteh AA, Koohshekan O, Khadivpour F, Kian M, Ghasemzadeh R, Pazoki M (2020) Industrial waste management using the rapid impact assessment matrix method for an industrial park. Global J Environ Sci Manage 6(2):261–274

Tchobanoglous G, Theisen H, Vigil S (1993a) Integrated solid waste management: engineering principles and management issues. McGraw-Hill

Tchobanoglous G, Theisen H, Eliassen R (1993b) Engineering principles and management issues. Mac Graw-Hill, New York, p 978

Vesilind PA, Worrell W, Reinhart D (2002) Municipal solid waste characteristics and quantities in solid waste engineering. Thomson Learning Inc., Singapore

Worrell E, Reuter M (eds) (2014). Handbook of recycling: state-of-the-art for practitioners, analysts, and scientists. Newnes

Zurbrugg C (2003) Solid waste management in developing countries. SWM introductory text on https://www.sanicon.net, p 5

Chapter 2
Landfilling

2.1 Introduction on Landfilling

Municipal Solid Waste (MSW) disposal has always been of considerable importance for governments worldwide. Dramatic growth in the municipal and industrial solid waste generation has been experienced in many countries due to extensive population size, rapid urbanization, changes in lifestyle patterns and acceleration in commercial and industrial developments in the past decades. To exemplify, waste production increased to a rate equal to 3 and 4.5% per year in Norway and the USA, between 1992 and 1996 respectively. Likewise, in Iran the rapid growth of MSW generation could be seen; for instance, between 2002 and 2009 waste generation increased by 7.3% and in 2006 a person produced about 746 gr of waste in a day. Landfilling has been considered an acceptable method for MSW disposal from an economical point of view.

Landfilling is the only waste disposal method that is able to deal with all materials in the solid waste stream. In biological or thermal treatment as an alternative, the process itself produces waste residues that should be landfilled subsequently (Tchobanoglous and Theissen 2005). Henceforth, in any solid waste management system, there will always be a need for landfilling. Moreover, landfilling is considered the simplest, and the cheapest in many cases, amongst the available disposal methods, so for the majority of solid waste disposal, it has been relied on throughout the human being history. In spite of the increase in land prices and environmental pressures leads to difficulties in finding suitable landing sites, landfilling is still the principal waste disposal method in several European countries, the USA and virtually developing world (Tchobanoglous and Theissen 2005).

An important component of integrated waste management is the safe and reliable disposal of municipal solid waste (MSW) and solid waste residues. Solid waste residues are constituents of the waste that remain after processing at a materials recovery facility, or after the conversion products and/or energy are recovered. Solid waste has been placed on or in the surface soils of the earth or deposited in the oceans

© Springer Nature Switzerland AG 2020
M. Pazoki and R. Ghasemzadeh, *Municipal Landfill Leachate Management*,
Environmental Science and Engineering,
https://doi.org/10.1007/978-3-030-50212-6_2

since ancient times (Ghasemzade and Pazoki 2017). Dumping of municipal solid waste in the ocean was officially ceased in the United States in 1933. The physical equipment used for the disposal of solid wastes and solid waste residuals in the surface soils of the earth is described as the term landfill. The application of landfills, in one form or another, has been the most economical and environmentally acceptable method for the disposal of solid wastes, after the dawn of the last century all around the world. Nowadays, the planning, design, operation, environmental monitoring, closure, and post-closure control of landfills are included in landfill management (Tchobanoglous et al. 1993).

In the USA the placement of solid waste on land is called a dump and in Great Britain, it is called a tip (as in tipping). The least expensive method for solid waste disposal in the open dump, so for almost all inland communities it was the original method of choice. The operation of a dump is straightforward as it involves just making sure that the trucks empty at the proper spot (Tchobanoglous and Theissen 2005; Tchobanoglous et al. 1993).

Prior to the enactment of the Resource Conservation and Recovery Act (RCRA) (1976), Americans had referred to typically not much more than open dumps as 'landfills'. In contrast to deterring vectors, there was no requirement for a daily layer of soil or preventing other hazards or nuisance conditions. Consequently, pre-RCRA facilities were commonly infested by insects and rodents and there were frequent fires. Protective liners were not typically constructed for these facilities; thus there is readily leaching into subsurface formations, including those which store groundwater. Regardless of subsurface geology or groundwater features, many landfills were sited at locations thought to be convenient. No impermeable substrate below the landfill unit was required for the prevention of liquids migration (Shayesteh et al. 2020; Wagner 1999).

If the final resting place for waste is an uncontrolled dump, no amount of careful waste collection or treatment will decline the threats to health or the environment from the disposal. An essential issue is the development of disposal sites away from open dumping.

Processes have changed dramatically so that landfills recently constructed, have overcome the problems previously associated with "dumps," during the past 20 years (Tchobanoglous et al. 1993) (Fig. 2.1).

2.2 History of Solid Waste Disposal

By the 1900s, "land disposal" of solid wastes was merely direct dumping on to the land surface followed by abandonment. "Nuisance areas," i.e. wetlands, on the periphery of many cities, were filled using layers of household waste and ash. However, in the early years of the century, the farsighted disposal methods of sanitary landfills started to evolve. In the United States, plain burying was applied. A precursor of today's modern sanitary landfill, opened in 1935 in California, was the first excavated site periodically covered with soil. In addition to MSW, industrial

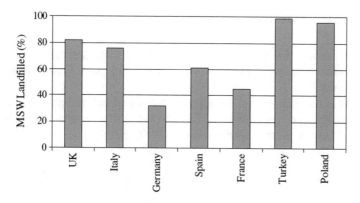

Fig. 2.1 Municipal solid waste landfilled in selected European countries

wasted was accepted in the landfill. Consequently, due to its content of hazardous materials, the site has secured a ranking on the U.S. EPA Superfund list (Greenberg and Hughes 1992).

The first proto-sanitary landfills were operated by Minoans (3000–1000 B.C.E.), placing their wastes, covered periodically with layers of soil, into large pits (Fig. 2.2).

Modern sanitary landfills have to meet strict obligations for siting, construction, operation and maintenance, and final closure according to RCRA laws. All active (i.e., receiving waste) municipal solid waste (MSW) landfills and not the stopped accepting MSW before October 1991 are subjected to RCRA regulations. The federal requirement for installing groundwater monitoring systems was combined in over a period of 5 years, due to the required complex technology. The nearest landfills to groundwater sources must comply before the farther away ones so that the drinking water sources could be protected. By April 9, 1994, it was required for landfill owners and operators to be able to pay the costs of closure, post-closure care, and clean-up of any identified releases (Shayesteh et al. 2020; Chambers and McCullough 1995).

Fig. 2.2 Sanitary landfill

2.3 Terms Definition

Figure 2.3 illustrates the general features of a sanitary landfill. Some commonly used terms to describe the landfill elements are defined as follows (Tchobanoglous and Theissen 2005).

The cell is the term used to describe the volume of material placed in a landfill during one operation period, usually 1 day (see Fig. 2.3b). The deposited solid waste and the daily cover material surrounding it is included in a cell.

Fig. 2.3 Cutaway views of a sanitary landfill **a** after geomembrane liner has been installed over compacted clay layer and before drainage and soil protective layers have been installed; **b** after two lifts of solid waste have been completed, and **c** completed landfill with landfill processes

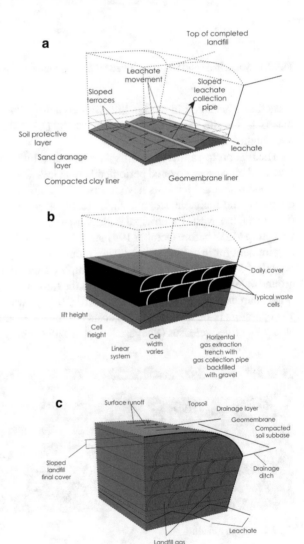

Usually, 6–12 Inch of native soil or other materials like compost, foundry sand, or auto shredder fluff used in the working faces of the landfill at the end of each running period, are contained in a *Daily cover*. From the past, the daily cover was employed for preventing rats, flies, and other disease vectors not to enter or exit the landfill. Today, daily landfill cover acts primarily in controlling the blowing of waste materials, to reduce odors and the entrance of water into the landfill during the process.

Over the active area of the landfill, there is a complete layer of cells, which is called a *lift*. Typically a series of lifts are comprised in a landfill.

Where the landfill's height goes beyond 50–75 ft, typically a *bench (or terrace)* is used. For several purposed, such as placement of surface drainage channel and location of landfill gas recovery piping, maintenance of the slope stability of the landfill is vital which is made possible using benches. The landfill cover layer is the final lift included.

The materials (both natural and artificial) used for lining the bottom area and below-grade sides of a landfill (see Fig. 2.3a) are called *Landfill liners*. Landfill leachate and landfill gas migration are prevented through successive layers of compacted clay and/or geosynthetic material consisted of liners.

After completion of all landfilling operations, the *final landfill cover layer* is placed over the entire landfill surface (see Fig. 2.3c). Landfill covers include successive layers made of compacted clay and/or geosynthetic material which is designed for stopping the migration of landfill gas and limiting the entrance of surface water into the landfill.

Leachate is a liquid formed at the bottom of a landfill. Generally speaking, percolation of precipitation, uncontrolled runoff, and irrigation of water into the landfill, resulting in the formation of leachate. Also, the water initially contained in the waste will consist of the leachate. Various chemicals from the solubilization of deposited materials in the landfill and the products of the chemical and biochemical reactions taking place within the landfill are contained in the leachate.

The mixture of gases within a landfill is called *Landfill gas*. Methane (CH_4) and carbon dioxide (CO_2) are the major constituents of the landfill gas which are the principal products of anaerobic biological decomposition of the biodegradable organic fraction of the MSW in the landfill.

The activities connected to the collection and analysis of water and air samples are called Environmental monitoring which is used to observe the movement of landfill gases and leaching materials at the landfill location.

The term used to express the necessary steps to close and secure a landfill site after the completion of the filling operation is Landfill closure.

The long-term maintenance of the completed landfill (typically 30–50 years) associated activities are referred to as Post-closure care.

The required actions to cease and clean up unexpected contaminants released to the environment are called Remediation.

The composition of refuse is approximately near 75 or 80% organic matters which constitutes mainly proteins, lipids, carbohydrate (cellulose and hemicelluloses), and lignin. As approximately two-thirds of this matter is biodegradable, the other one-third fraction is unbiodegradable (Fig. 2.4) (Ghasemzadeh et al. 2017).

The biodegradable fraction of the refuse can be divided into two portions; i.e. the readily biodegradable portion (food and garden wastes) and a moderately biodegradable one (paper, textiles, and wood). The major biodegradation pathways for the decomposition of most of the organic materials in solid waste are summarized in Fig. 2.5 Due to the heterogeneity of waste and landfill operating features, the landfill has a sort of diverse ecosystem. This phenomenon increases the stability of the ecosystem; the system is dramatically affected by the environmental condition (such

Fig. 2.4 Total MSW generation by category in 2008

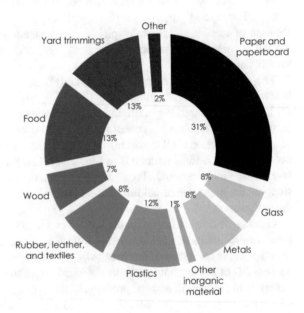

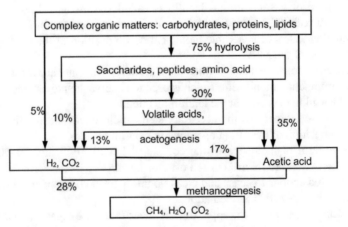

Fig. 2.5 Predominant decomposition pathways for common organic waste constituents

Table 2.1 Important microbial groups promoting anaerobic waste degradation

Microbial group	Substrate
Amylolytic bacteria	Starches
Proteolytic bacteria	Proteins
Cellulolytic bacteria	Cellulose
Hemicelluloytic bacteria	Hemicellulose
Hydrogen-oxidizing methanogenic bacteria	Hydrogen
Acetoclastic methanogenic bacteria	Acetic acid
Sulfate-reducing bacteria	Sulfate

as temperature, pH, presence of toxins, moisture content, and the oxidation–reduction potential), however (Ghasemzadeh et al. 2017). Electron donors, primarily organic ones, are predominant in the landfill environment. Carbon dioxide and sulfate are the major electron receptors. There are seven significant physiological microbial groups that take part in rate-determining steps of fermentation and methanogenesis which are listed in Table 2.1.

Some of the landfill research studies6 have stated that five consecutive and definite phases are included in the process of stabilization of waste. Not only the rate and features of the produced leachate and gas evolve from a landfill are different, but also determine the microbial mediated processes which occur inside the landfill during these phases. In the following section, the phases experienced by degrading wastes, are described. In Table 2.2 the leachate characteristics during the waste degradation phases are outlined.

Phase 1: Initial Adjustment Phase
Initial placement of solid waste and aggregation of moisture in the landfills is related to this phase. Until enough moisture forms and supports an active microbial community, an acclimation period (or initial lag time) is detected. Introductory variations in environmental constituents occur so that favorable conditions for biochemical deposition are created.

Table 2.2 Landfill constituent concentration ranges as a function of the degree of landfill stabilization (Vesilind et al. 2002a)

Parameter	Transition	Acid formation	Methane formation	Maturation
Chemical oxygen demand, mg/l	4800–18,000	1500–71,000	580–9760	31–900
Total volatile acids, mg/l as acetic acid	100–3000	3000–18,800	250–4000	0
Ammonia, mg/l-N	120–125	2–1030	6–430	6–430
pH	6.7	4.7–7.7	6.3–8.8	7.1–8.8
Conductivity, μS/cm	2450–3310	1600–17,100	2900–7700	1400–4500

Phase 2: Transition Phase

Depletion of oxygen trapped within the landfill media is evidence of a transformation from an aerobic to an anaerobic environment which a result of the exceedance of the field capacity in the transition phase. Shifting of electron acceptors from oxygen to nitrates and sulfates and the displacement of oxygen by carbon dioxide is accompanied by a trend toward reducing conditions.

Measurable concentrations of chemical oxygen demand (COD) and volatile organic acids (VOAs) can be monitored in the leachate in the ultimate of this phase.

Phase 3: Acid Formation Phase

The formation of intermediate volatile organic acids at high concentrations throughout this phase is a consequence of the continuous hydrolysis (solubilization) of solid waste, followed by (or concomitant with) the microbial conversion of biodegradable organic content. Mobilization of the metal species is followed by a reduction in the pH values. The main features of this phase are reasonable biomass growth related to the acid former (acidogenic bacteria), and the rapid consumption of substrate and nutrients.

Phase 4: Methane Fermentation Phase

Methane-forming consortia (methanogenic bacteria) consume intermediate acids and convert them into methane and carbon dioxide during phase IV. Sulfate and nitrate are reduced to sulfides and ammonia respectively. The growth of methanogenic bacteria is supported by the elevation of the pH value which is controlled by the bicarbonate buffering system. Complexation and precipitation result in the removal of heavy metals from the leachate.

Phase 5: Maturation Phase

During the final state of landfill stabilization, due to the lack of nutrients and available substrate, the biological activity becomes relatively inactive. Gas production significantly decreases, and leachate strength remains steady at much lower concentrations. Oxygen and oxidized species may gradually reappear. Production of humic-like substances may be accompanied by the slow degradation of resistant organic fractions, however. The movement to final stabilization of landfill solid waste is subject to the physical, chemical, and biological factors within the landfill environment, the age, and characteristics of landfilled waste, the applied operational and management controls, as well as the site-specific external conditions.

2.4   Leachate and Gas Generation

The classification system adopted by the state of California in 1984 is perhaps the most widely accepted classification system for landfills, despite the fact that many other landfill classification systems have been devised over the years. There are three classifications used in the California system, as reported in the following table.

Class	Type of waste
I	Hazardous waste
II	Designated waste
III	Municipal solid waste (MSW)

Most of the landfills across the United States are designed for commingled MSW. Many of these Class III landfills accept small amounts of nonhazardous industrial wastes and sludge from water and wastewater treatment plants. In many states, dried sludge of treatment plants with a solid content of 51% or greater without any free-flowing liquid content, are accepted. Recently, no liquid waste is accepted into MSW landfills due to federal regulations.

In several locations throughout the United States, an alternative method of land-filling is tried, which involves shredding of the solid wastes before landfilling. An exemption from daily cover requirements is put on the shredded (or milled) waste of up to 35% greater density than unshredded waste.

Blowing litter, odors, flies and rats have not been considerable problems. When faster waste decomposition is an operating goal, the application of shredders may be necessary despite the typical decline of using shredders.

Baling the MSW for placement in the landfill is another method. The required equipment for compaction is eliminated in this approach and it also takes advantage of easier handling. At a production facility places at an external transfer station or at an unloading station of the landfill property, the bales are made. Forklifts or other similar equipment is used to move and stack the bales to the working face on flatbed vehicles. Just as a lift is completed, covering is done, although daily covering is not always required. Some of the nonhazardous wastes may possibly release substances in concentrations that exceed the acceptable water quality regulations established by different state and federal institutes, which are called designated wastes. Usually identified designated wastes, i.e. combustion ash, asbestos, and other similar wastes are typically isolated from materials placed in municipal landfills by means of being placed in lined monofills.

2.5 Natural Attenuation Landfills

In non-limited sites, when there is little or no engineering of the boundaries, leachate [the liquid formed within a landfill site which is composed of the liquids entering the site (including rainwater) and the leached material from the wastes percolating downwards through the waste] migrates into the surrounding environment. This phenomenon is referred to as the dilute and attenuates principle. Some biological and physicochemical processes are involved in attenuation of the leachate both within the waste and in the surrounding geology. Studies in the mid-1970s showed attenuation of leachate moving through different unsaturated layers. Moreover, it showed that the attenuation processes (defined to include dilution) can be employed in the treatment of the landfill leachate migrating from the site. In "dilute and attenuate" landfill, no leachate collection, and treatment facility are required, as the formed liquids move from the base; moreover there is no necessity for expensive landfill lining/engineering. These are the advantages of the method. The risk of migration of leachate reduces through dilution within groundwater, but naturally, that groundwater is contaminated. The mentioned risk is deemed acceptable to suppose the "dilute and attenuate" principle to be effective (Bagchi 2004; Pazoki et al. 2015a).

2.6   Containment Landfills

A much greater degree of site design, engineering and management and some sort of control over the hazardous waste disposal to landfill is required in the containment principle of landfill. In spite of the greater need for engineering to achieve containment and considerable variables involved in the management of water and other parameters, it is now the accepted means of inland disposal in the developed nations. Prevention of the leachate generated within the waste from migrating beyond the site boundary is the underlying principle of containment landfill. A landfill site, in which the rate of leachate released into the environment is considerably low, is a "Containment site". Biodegradation and attenuating processes are allowed to occur through retaining the polluting constituents of wasted within landfills for enough time; consequently, the migration of polluting species with an unacceptable concentration is prevented. Also, the release of gases may be dramatically reduced, according to site engineering. The key phrases in the just stated definition are the rate of release is extremely low and at an unacceptable concentration, implies possible migration of some leachate; however, the contributing risk is acceptable. Release of leachate at even the most highly engineered containment landfills must be expected at some time in the future. Regarding this issue, the operation and management of landfills should be considered deeply to keep any release at an acceptable concentration. The containment landfill risk assessment has been explained elsewhere. The concepts of a sustainable landfill and fail-safe landfill16 have been devised with an acceptable and managed risk in mind (Vesilind et al. 2002a; Allen 2001).

2.7   Bioreactor Landfill

The aim of landfilling has changed from the storage of waste to the treatment of waste with a bioreactor landfill. An isolated system from the environment enhancing the degradation of waste by microorganisms is a bioreactor landfill. Addition of specific elements (nutrients, oxygen, or moisture) and controlling other elements (such as temperature or pH) promote microbial degradation. Landfills with more solid waste are most widely created by this method. The working life of the landfill is increased

Fig. 2.6 Bioreactor landfill using leachate recirculation

and the construction of new landfills is delayed with the potential to add more waste to the landfill.

2.7.1 Bioreactor Landfill Definitions

A bioreactor landfill is a landfill or landfill cell which is controlled through actively managing the liquid and gas conditions to increase or enhance the stabilization of waste. The degree of organic waste decomposition, rates of conversion and process efficiency is considerably increased over other types of landfills in the bioreactor landfill (Solid Waste Association of North America (SWANA)). Therefore, a bioreactor landfill is: "a MSW landfill or a portion of a MSW landfill in which, in order to reach a minimum average humidity content of at least 40% by weight, liquids except for leachate or landfill gas condensate, are added in a controlled manner into the waste mass (often in combination with recirculation leachate) to accelerate the anaerobic biodegradation of the waste (Vesilind et al. 2002a; Kumar et al. 2011) (Fig. 2.6).

2.7.2 Bioreactor Landfill Fundamentals

A bioreactor landfill is usually defined as a sanitary landfill in which enhanced microbiological processes are used to convert and stabilize the most likely and moderately decomposable organic waste components in a short time (typically 5–10 years) compared to a common landfill (typically 30–50 years or more). The waste stabilization process is controlled, monitored and optimized by a bioreactor landfill operator rather than be simply contained as required under current regulations. In comparison to standard methods, bioreactor landfills provide a more sustainable and green waste management strategy, if operated in a controlled and safe way. The stabilization process is improved using specific system design and operation modifications which are required in the bioreactor landfill. These include Liquid Addition: A desirable environment for organisms responsible for waste decomposition is prepared by the addition of moisture to landfilled waste. The microorganisms usually need more moisture than the amount available in the waste, so liquids are added to the landfill waste through certain design and operational modifications. The most common source of liquid is the recirculation of leachate, but other moisture supplies can also be used. Air Addition: Another important feature for some bioreactor landfills is air addition. The aerobic stabilization of the landfilled waste is enhanced through the addition of air and consequently oxygen. The same process is used in a traditional waste compost system to decompose waste. As compared to anaerobic waste decomposition, aerobic waste decomposition is a faster process. In the cold area, the aerobic technique is possibly helpful for constructing bioreactor technology. Major components of an anaerobic bioreactor landfill including leachate/moisture addition system, gas recovery system, bottom-line system, and leachate collection system are depicted in the schematic diagram (Vesilind et al. 2002a; Pohland 1996).

Other Factors: The primary technologies for enhancing waste stabilization in controlled bioreactor landfills are moisture addition, and of the second importance, air addition, however, other environmental factors are sometimes suggested for controlling. These factors include temperature, pH, and nutrient level. For the mesophilic microorganisms, the optimum temperature is in the range of 34–40 °C while for thermopiles microorganisms, it is up to 70 °C. In cold regions, aerating landfilled waste is employed to warm the system in the preliminary phase of an anaerobic bioreactor, since low temperatures may cause problems. Temperature control is a critical issue in operating an aerobic bioreactor due to the possibility of ignition. The activity of methane forming bacteria is influenced by pH. The optimum pH range is known to be 6.8–7.4 for the methane forming bacteria (Vesilind et al. 2002b).

2.7.3 Classification

Depending on the studies, context, or regulatory programs, a bioreactor landfill may be defined in various manner. For the purpose of this text, the bioreactor landfill is a landfill designed and functioned to perform the controlled treatment (stabilization) of the solid reuse. "What the difference between a bioreactor landfill and a landfill that carries out the leachate recirculation is" a typical question. From the point of view of this guide (not necessarily from a regulatory perspective), a landfill recirculating leachate in a controlled process to improve the stability of waste is a bioreactor. If the only goal of leachate recirculation in a landfill is to dispose of it, the landfill is not considered as a bioreactor (though many of the same design and operation issues apply). In the following sections, the other terms describing specific types of bioreactors are presented (Vesilind et al. 2002b).

2.7.4 Benefits

Many potential advantages over a conventional landfill are offered when a landfill is operated as a bioreactor. Waste Stabilization: The bioreactor waste management operation will be more sustainable through accelerated waste stabilization which is the primary benefit of operating a bioreactor landfill. This is related to several other advantages.

2.8 Structural Characteristics

The solid waste behaves in a quite similar manner to other fill materials as it is first placed in a landfill. The apparent repose angle of the placed waste material in a landfill is about 1.5:1. Since solid waste is likely to slip in too steep slope angles, the used slopes for completed parts of a landfill are from 2.5:1 to 4:1; 3:1 is the most typical angle. The slope stability needs to be considered more carefully, as the landfills increase in size. Waste has placed at significantly sharp slopes and at heights in excess of 100 ft. due to uncontrolled dumping, in some cases. At a number of large uncontrolled sites, significant loss of life has occurred because of major slope problems. Since 1980, a major slope failure has happened every few years in the United States. A slope stability analysis is required as part of the design process in order to prevent slope failure (Vesilind et al. 2002a, b).

Slope Stability. When more waste placed in the landfill than had been previously predicted, problems have often happened. Either foundation-type or slope failure may occur due to this issue. When limited undersurface movement begins to take place, cracks may open within the waste, which is a sign of the possible failure occurrence. Careful monitoring of the cracks is necessary and filling the cracks with materials followed by their reappearance, is a signal of immediate failure danger. Soil and waste mechanics analysis is used to determine the slope stability of a landfill. At any site where the slope of the sites goes beyond 3:1, this would normally take place. The arrangement of waste placement, the angle of repose of the waste, the stress–strain characteristics of the liner and cover materials, and the ability of the foundation soils to support the landfill are considered in the mentioned analysis. Where the height of the landfill exceeds 50 ft., many landfills are benched due to the slippage problems. Furthermore, surface water drainage channels are placed using benches and landfill gas recovery piping is located (Fig. 2.7).

Fig. 2.7 Landfill slope failure

2.8.1 Seismic Protection

The seismic protection of landfills is connected to the slope stability analysis. Determining the critical design factors required to prevent slope failure during an earthquake using a seismic analysis, is of considerable importance in seismically active areas. Although the possibility of this sort of failure is less than the previously described slope stability failures, it has occurred. Slope failure in a portion of landfill or displacement of the cover materials from the waste material will occur due to transmitted ground motion through the waste which is related to seismic failures. Specialized technical skills are needed in this seismic analysis.

2.8.2 Settlement of Landfills

The settlement of landfills occurs due to the process of decomposition of the organic materials and the resulting weight loss through the extrusion of the landfill gas and leachate components. An increase in overburden mass as the addition of landfill lifts and percolation of water into and out of the landfill also results in a settlement. Break downs of the landfill surface and cover ruptures, and misalignments of gas recovery facilities, cracking of manholes and interference with subsequent use of the landfill after closure, are landfill settlement consequences.

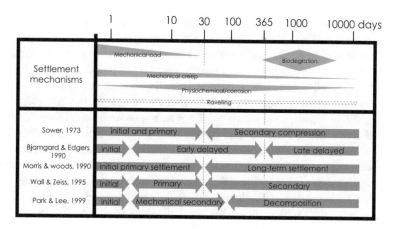

Fig. 2.8 Occurrence of settlement mechanisms and temporal classifications adopted by selected publications

2.8.3 *Mechanisms of Settlement and Temporal Classification*

Overview of mechanisms of settlement and temporal classification shows in Fig. 2.8 (Zekkos and Flanagan 2011; Sahadewa et al. 2011).

2.8.4 *Influencing Factors*

Mechanisms and factors influencing landfill settlement show in Fig. 2.9 (Zekkos and Flanagan 2011).

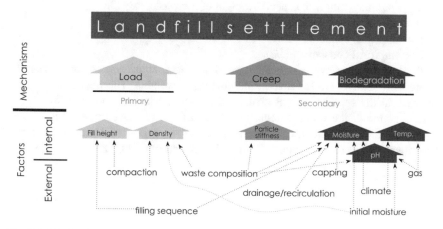

Fig. 2.9 Mechanisms and factors influencing landfill settlement

Because of some factors including rough settlement characteristics, different bearing capacity of the upper layers of the landfill, and the potential problems resulting from gas migration, even with the use of gas collection facilities, generally making permanent facilities on completed landfills is not advised. Control over the deposition of specific materials within the landfill operation is possible at the time the final application of the landfill is specified before the start of the waste placement. To exemplify, in locations, in which buildings and/or other physical facilities are to be placed in the future, materials like construction and demolition wastes which are moderately inert can be placed. The placement of structures on completed landfills is limited according to recent regulatory trends. A combination of processes including loading and biodegradation-related phenomena is involved in landfill settlement, as it is a well-known principle. Whilst biodegradation is not sufficiently dealt with existing geotechnical settlement models, load effects are handles effectively.

2.8.5   Effect of Overburden Pressure (Height)

The average specific weight of waste in a lift depends on the depth of the lift because the density of the placed material in the landfill will experience a rise with the weight of the material placed above it. In an overburdened landfill, the maximum specific weight of solid waste residue will range from 1750 to 150 lb/yd3. An estimation of the increase in the specific weight of the waste as a function of the overburden pressure can be obtained using the following relationship:

$$D_{W_p} = D_{W_i} + \frac{p}{a + bp}$$

where D_{Wp} = specific weight of the landfill material at pressure p, lb/yd3.
   D_{W_i} = initial compacted specific weight of the waste, lb/yd3.
   p = overburden pressure, lb/ft^2.
   a = empirical constant, yd3/ft^2.
   b = empirical constant, yd3/lb.
   0 represents the typical specific weights versus applied pressure curves for compacted solid waste for several initial specific weights. The importance of the increase in the specific weight of the waste material in the landfill is due to (1) determination of the actual amount of waste that can be placed in a landfill up to a given grade limitation and (2) determination of the degree of settlement that can be expected in a completed landfill after closure (Pfeffer 1992; Edgers et al. 1992) (Fig. 2.10).

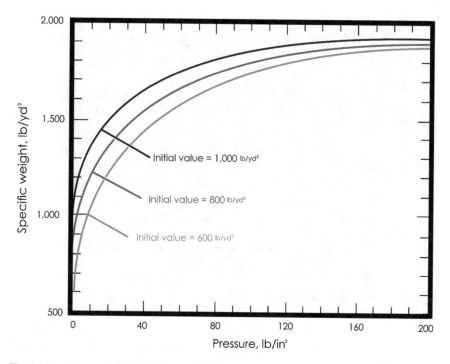

Fig. 2.10 Specific weight of solid waste placed in landfill as a function of the initial compacted specific weight of the waste and the overburden pressure

2.8.6 *Extent of Settlement*

The initial compaction, the characteristics of wastes, the degree of decomposition, the height of the completed fill and the effects of consolidation when water and air are forced out of the compacted solid waste, are the key factors in the extent of settlement. 0 shows some data on the degree of settlement to be expected in a landfill as a function of the initial compaction. Various studies show that about 90% of the final settlement occurs within the first 5 years (Fig. 2.11).

Prediction of settlement rates in landfills that account for waste decomposition, in addition to the weight of the material, is carried out using several proposed methods (Edgers et al. 1992; Edil et al. 1990; El-Fadel et al. 1999; Watts and Charles 1999). The following is a description of a one-dimensional model for landfill settlement:

Intermediate secondary settlement:

$$S_{(t)} = H_0 C_{\alpha 1} \log \left(\frac{t}{t_{initial}} \right) t_{initial} < t < t_2$$

Long-term secondary settlement:

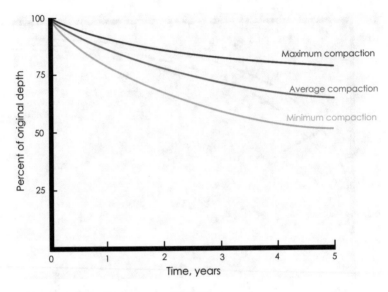

Fig. 2.11 Surface settlement of compacted landfills

$$S_{(t)} = H_0 C_{\alpha 2} \log\left(\frac{t}{t_2}\right) t_2 < t < t_{initial}$$

where $C_{\alpha 1}$ = coefficient of intermediate secondary compression (varies from 0.015 to 0.035).

$C_{\alpha 2}$ = coefficient of long-term secondary compression (varies from 0.132 to 0.25).

$t_{initial}$ = end of initial settlement period (16 days).

t_{final} = end of field experiment observations (1576 days).

t_2 = time at which slope of stress–strain curve changes (day).

2.9 Planning

The goals must be planned according to a long-term time scale, while the solid waste disposal plan is set. Planning of a ten-year time frame is taken as a short-term planning. After thirty years, the anticipation of solid waste generation and new disposal technologies would be difficult. Thus a thirty-year time frame looks appropriate. The establishment of the necessities of a landfill site is the first step in planning for a new landfill. Any ancillary solid waste functions, such as leachate treatment, landfill gas management, and special waste services (i.e., tires, bulky items, household hazardous wastes) must be supported by the landfill capacity provided for the selected design period. Requirements for the handling of recyclable materials (material recovery facilities) and green waste compost is prepared in some sites.

The disposal facilities should be estimated prior to the determination of landfill capacity. Sufficient service life is not provided by a too-small landfill and the expense of its construction will not justify. On the other hand, a huge landfill may result in high up-front capital costs, consequently prohibiting the construction of other required public facilities and may eliminate many potential sites. A historical guide of disposal quantities is provided by records from the past several years. If scales were applied at existing landfills, this information can be very precise in some cases. But is the amount of delivered solid waste to the landfills is only an estimation, the information is doubtful. A per capita disposal rate can be calculated, using the historical population. The steady rates of per capita solid waste generation in the U.S. were about 4.5 lb per person per day, since 1990. On the other hand, per capita disposal has decreased from 3.12 lb per person per day to 2.43 lb per person per day. The amount of waste landfilled, annually has remained almost constant since 1990, ranging from 134 to 142 million tons per year. In-place density (the density once the waste has been compacted in the ground) should be predicted if possible. Determination of density can be easily done by conducting routine aerial surveys of the landfill and afterward calculation of the volumes for an existing available landfill. The volume of cover material is included in the calculation in this method. About 20 to 50% of the volume of the landfill may be cover material if dirt is used as the daily and final cover. 1200 lb/yd3 (700 kg/m3) is a typical in-place density. Estimation of the compaction of the waste by individual refuse constituents is necessary when some materials from the solid waste are recovered and consequently, the compaction characteristics change markedly. Even with the difference of the actual compaction, bulk densities of the components can be applied in such calculations. Some bulk densities are listed in Table 2.3 that can be used for this purpose (Vesilind et al. 2002a).

To simplify the volume, mass and density calculations for mixed materials, a container holding a mixture of materials, each of which has its own bulk density, can be considered. The mass calculation for each contributing material can be performed through addition, and then divided by the total known volume as presented in the

Table 2.3 Bulk densities of some uncompacted refuse components

Microbial	g/cm^3	Lb/ft^3
Light ferrous (cans)	0.100	6.36
Aluminum	0.038	2.36
Glass	0.295	18.45
Miscellaneous paper	0.061	3.81
Newspaper	0.099	6.19
Plastics	0.037	2.37
Corrugated cardboard	0.030	1.87
Food waste	0.368	23.04
Yard waste	0.071	4.45
Rubber	0.238	14.9

below equation.

$$\frac{M_A + M_B}{\left[\frac{M_A}{\rho_A}\right] + \left[\frac{M_B}{\rho_B}\right]} = \rho(A + B)$$

where:
   A = bulk density of material A.
   B = bulk density of material B.
   V_A = volume of material A.
   V_B = volume of material B.
   M_A = mass of material A.
   M_B = mass of material B.
   ρ_A = bulk density of material A.
   ρ_B = bulk density of material B.

   In the presence of more than two different materials, the extension of this equation can be used. The equation for calculating the overall bulk density in the case of the expression of two materials at different densities in terms of their weight fraction is as follows.

   The extension of this equation can be used if more than two materials are involved. Important design and an operational variable is the volume reduction obtained in waste baling or landfill compaction. With the original volume V_O of solid waste and the final volume, after compaction, of V_C, the volume reduction is calculated according to the below equation:

$$\frac{V_C}{V_O} = F$$

where:
   F = fraction remaining of initial volume as a result of compaction.
   V_O = initial volume.
   V_C = compacted volume.

   As the constancy of mass (the same sample is compacted so there is no gain or loss of mass), volume and so volume reduction can be calculated by volume = mass/density as a result of compaction.

   The density may experience a change, over the life of a landfill. The volume necessary for the landfill is affected by the following factors:

– New Regulations. Waste diversion/recycling goals have been taken up in many states. Diversion rates of 70% and beyond may be obtained if the goals are completely applied.
– Competing Facilities. Some of the planned waste would be accepted by other present landfills, or solid waste may be imported due to closure of other existing landfills.

- Different Cover Options. Depending on the landfill size, 20 to 50% of the available landfill volume may be consumed by dirt used as daily and final cover at a landfill. The available volume for the solid waste will significantly rise if the landfill applies foam, tarps, or mulched green waste for cover.
- Nonresidential Waste Changes. All external waste, currently going to the landfill will be considered in a per capita generation rate. The nonresidential waste contributed to more than 50% of the landfill volume needed, in some cases. Large military, agriculture and manufacturing facilities, and cities with a large percentage of communities and tourists are some examples. The upcoming waste projections are considerably influenced by either closure or opening of these facilities.

2.10   Landfill Site Selection

Finding a convenient site is required when the landfill site has been determined. Imposing the least negative effects on the environment or on the population due to the sited location of the disposal facility is the main goal of the landfill site selection process. Identification of the best available disposal location meeting the standards of government regulations diminishing economic, environmental, health, and social costs is the result of a careful evaluation process for a sanitary landfill siting. Outcomes of the evaluation processes or methodologies can be approved and defended when the process is structured to make use of available information effectively, ensuring the reproducibility of the results obtained. One of the most important decisions made by a metropolis in devising its waste management plan is to choose a proper site for developing a landfill. High expenses on waste transport, site development, site operations, or environmental protection is imposed by an inappropriate selected site. Furthermore, future political problems may occur due to public opposition. It is necessary to follow a systematic process for the selection of a proper site. The priority of criteria for selection should be ordered according to local climatic, political, and cultural situations. Evaluation of potential locations for the long-term disposal of solid waste includes the considerable factors which follow:

- Haul distance
- Location restrictions
- Available land area
- Site access
- Soil conditions and topography
- Climatological conditions
- Surface-water hydrology
- Geologic and hydrogeologic conditions
- Existing land-use patterns
- Local environmental conditions
- Potential ultimate uses for the completed site.

Results of an intricate site survey, results of engineering design and economical studies, managing one or more environmental impact estimations, and the conclusion of public hearings are usually the foundation of the ultimate selection of a disposal site.

Figure 2.12 illustrates an overlay procedure to collect and present the pertinent site selection information (Tchobanoglous et al. 1993). Sites with a wide range of characteristics are provided by comparing the proposed site to an ideal site through a site scoring procedure. The input from the public can be used to prepare a list of technically plausible sites. Landfills are usually considered as LULUs (locally undesirable land uses). From this point of view, the landfill development process should include the public idea. While a landfill is to be sited, special procedures should be specified by state or local regulations to interact with the public. To successfully site a new landfill, it is often necessary to perform a simultaneous public information and negotiation process with the technical development activities.

Public participation should be a process of recommendation. The aims of a public participation program are (Tchobanoglous and Theissen 2005):

- Improving full public understanding of the need for a landfill and its operation principles
- keeping the public knowledgeable on the situation of various planning, design, and operation activities
- asking valuable opinions and perceptions about landfill development from concerned citizens
- Promotion of the public concern about their role as waste generators.
- A public participation program holds the advantages of:
- increasing the possibility of public approval for the final plans
- Provision of useful information to decision-makers, especially where not easily qualified issues or factors are involved
- ensuring that all issues are considered carefully
- identifying how to compromise with affected communities
- Provision of a safety control of a forum where accountability of decision-makers is aired through criticism
- convince decision-makers about the necessity to be responsive to issues beyond those of the immediate project in an effective mechanism
- encouraging the public to reflect on their role as waste producers effectively
- The negative effects of improperly managed public participation programs may include
- potential for confusion of the issues if too many new subjects are brought about
- possible gained erroneous information due to unknowledgeable participants
- addition of expenditures to the project due to public involvement
- possible delays in the project due to public involvement
- The appropriate people may not be involved in the effort, or the possibility of not developing interest by the citizens until it is too late for changes to be initiated
- The public may resist against landfilling despite the best efforts at public participation.

Fig. 2.12 Overlay maps of various site criteria Used in the screening of potential landfill sites

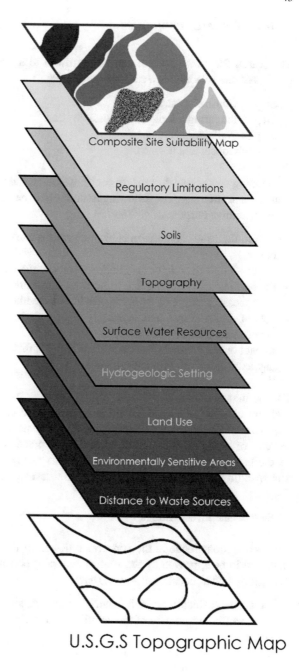

Composite Site Suitability Map

Regulatory Limitations

Soils

Topography

Surface Water Resources

Hydrogeologic Setting

Land Use

Environmentally Sensitive Areas

Distance to Waste Sources

U.S.G.S Topographic Map

2.10.1 General Principles

The search for a new landfill can be narrowed to a single step since the general process of site selection is a step-by-step sequence.

Step 1: Constraint mapping.
Step 2: Preparing a long list of possible sites.
Step 3: Walkover surveys.
Step 4: Conceptual designs.

Developing conceptual designs for each considered site is vital to ease the identification of the desired site. The following issues need to be designed to permit estimation for comparative purposes:

− the required site capacity (in cubic meters of waste) volume of the daily and final cover
− the needed resources for installation of an adequate leachate control system
− the required extent of surface water diversion works
− the necessary extent of works to provide all-weather access to the site
− cost of all the above, and separately identified, the additional cost of importing cover material from elsewhere
− impact on the waste collection system of using this site and the cost of providing any extra resources (vehicles, transfer stations).

For the first purchase of the land, estimation of costs is necessary for this service and any ultimate sale on completion of productive use should be evaluated. The approximate lifetime over which the initial development costs could be amortized and the cost of supplying cover material would be spread is indicated by the estimation of the capacity of each site. Indicative "costs per cubic meter of waste" for the site-specific former considered elements are developed through estimation (Table 2.4).

Step 5: Site investigations

The assumptions made in the conceptual design of the foregoing issues should be designed to be approved by the studies generally performed taking advantage of a drilling rig and a mechanical shovel (backhoe):

− the available quantity of soil material within the site for cover purposes
− permeability of the base of the landfill and of the material to be used for the final cover
− bearing capacity of the base of the landfill
− stability of any slopes to be cut
− groundwater regime
− quality of ground and surface water baseline.

Table 2.4 Checklist for walkover survey

Transport Aspects
A.1 To what point is all-weather access presently available?
A.2 How long does it take to travel from the urban area to the nearest accessible point to the site?
A.3 How far (by new or upgraded road) is the site from this point?
A.4 Will vehicles are able to gain access to all parts of the site (via site roads)?
A.5 Will access be unusually expensive to provide (large or long embankments, bridges, cuttings)?
Natural Features
B. Is the site presently well drained?
B.2 Are there established watercourses within or adjacent to the site?
B.3 Is there evidence of any ephemeral streams, springs, or sinkholes?
B.4 Can a high water table be inferred from the vegetation anywhere on the site?
B.5 Are surface water diversions likely to be extensive (considering extent of catchment)?
B.6 From knowledge of the geology of the area, does the morphology of the site suggest significant or minimum depths of soft material (for daily cover and other purposes)?
B.7 Are there areas within a few kilometers of the site which may be suitable for borrow material?
B.8 Is there any evidence of geological features on or near the site?
B.9 Are there any features that will significantly limit the useful area of the site for landfilling?
Land Use
C.1 What is the present land use of the site and the route of any access road to it?
C.2 What is the present land use in the immediate vicinity of the site and access route?
C.3 Are there likely to be any water abstractions (for drinking or livestock watering) downstream of the site (for example, within I km)?
C.4 Are there any overhead power lines crossing the site?
C.5 Is there evidence of buried electrical cables or water pipes on the site?
C.6 Is there any evidence to suggest where the nearest point of water distribution or electricity distribution network might be to the site?
C.7 Are there any places of historic or cultural significance nearby?
C.8 Is there likely to be a need for resettlement?
Public Acceptability
D.1 Are there any significant population centers on the principal route to the site which will be adversely affected by increased traffic volumes?
D.2 Is the site overlooked by, or overlooking residential or commercial development, or socio-politically sensitive sites?
D.3 Where are the nearest inhabited dwellings (e.g., farms)?

The success of the sitting and design of the landfill is critically dependent on these site investigations. An experienced geotechnical engineer followed by a hydrogeologist should interpret the results of designed and supervised investigations.

Step 6: Feasibility report, including environmental impact assessment

The viability of developing the desired site as an efficiently managed landfill site is assessed by a feasibility report provided by the waste management staff of the metropolis, with one preferred site have identified. The plausibility of the development in five areas needs to be clearly demonstrated in the report:

– physical and environmental
– technical
– economic
– social and cultural
– legal.

The summary of the selection process and the basis of the identification of the preferred site should be justified at the beginning of the report.

The current costs of waste disposal at the nearly closing existing site(s), the expected cost of the design, construction, operation, completion, and aftercare of the new sanitary landfill should be considered in the economic justification of the site.

– closing down the existing site(s)
– the immediate costs of gathering and transporting wastes to the existing site(s)
– the costs expected in collection and transportation of wastes to the new sanitary landfill
– any proposed apportionment of costs between operating departments or benefiting cities
– any changes in cost recovery which may be expected during the lifetime of the new sanitary landfill.

The implications on long-term budgets can be obviously seen if the mentioned cost analyses are expressed as cash flow forecasts and in terms of net present value (NPV) as well.

Step 7: Final decision

Obtaining support, if not secured yet, from the proper committee of the municipality, regulatory authorities, and the provider(s) of the funds for the landfill development, is the ultimate obstacle prior to carrying out any project.

The upcoming issues should be clearly explained by the promoter of the scheme, in order to convince the sponsor even though the feasibility report is the primary supporting document:

- there is an immediate necessity for the promotion of current waste disposal practices
- the most appropriate way will be by developing and operating a long-term landfill
- the best available site in the area is the selected site
- satisfactory environmental standards will be met by the operation of new landfill
- the impact of its introduction on the waste collection system has been completely considered
- the cost of the changes to the waste management system is reasonable and affordable.

Also, various benefits can be obtained using geographical information systems (GIS) and related software such as "Arc view". Making a decision will be more straightforward using the graphical output of this software.

2.10.2 Criteria for Landfill Siting

The site selection of landfills is performed on the basis of number of criteria. Environmental, political, financial and economic, hydrologic and hydrogeological, topographical and geological are the criteria involved. The availability of construction material and other parameters are also included. In the next sections, each criterion will be briefly explained.

2.10.2.1 Environmental Criteria

1. **Ecological value of the flora and fauna**

The present vegetation and fauna may be destroyed through direct and indirect use of a landfill. Careful evaluation of the actual vegetation and fauna of the considered area is a necessary step for making a decision.

Biodiversity, naturalness and other characteristic features are the foundation of evaluation of ecological value. Noise creation in the surroundings due to landfill activities is an example of indirect use.

2. **Odor and dust nuisance**

The release dust and odor is the reason why a new landfill should not be located within a distance of a housing area. Sensing dust and odor is prevented through the determination of the necessary safe distance of a landfill, which is dependent on one local wind direction and speed. Furthermore, appropriate soil cover can minimize the problems of odor and dust.

3. **Nuisance by the traffic generation**

More traffic jams will be generated due to the construction of a new landfill. Distance to the collection area, means of transportation and application of transfer stations, determine the amount of additional traffic. The degree botheration caused by access roads passing through housing areas is much more than access roads through the open countryside. The noise impacts of vehicles related to landfill are decreased through the usage of routing vehicle traffic through industrial, commercial or low-density population areas.

4. Risks for explosion or fire

Landfill gas makes a chance of explosion and/or fire. A barrier to prevent the spreading of fires and smoothing fires are the function of soil cover. Minimization of the dumping of flammable loads can decrease fire risk if the incoming trucks are properly policed.

5. Other nuisance for neighboring area

The vermin attracted by the organic parts of the waste of the landfill (rats, mice, birds, and insects), clutter blown by the wind, construction, compaction or trucks noise are the other kind of nuisance. The nuisance developed vermin, is solved by the daily cover. Development of idle pools of water which causes breeding of mosquitoes is prevented through regular grading of soil cover to fill in low spots.

6. Ecological, scientific or historical areas

Locating of a landfill is not reasonable in especially national parks and natural protected areas and also historical areas.

7. Tourist/recreation areas

An existing recreational area or its vicinity is not a suitable place for a new landfill. However, in some kinds of recreation areas like car/motor racing, a landfill is feasible. Moreover, a landfill can be finally used as a recreational area.

2.10.2.2   Political Criteria

1. Acceptance by the local municipalities

Occasionally, the planned sites are located in different regions, because the political acceptance of a new landfill location can differ in each region. The eagerness of the domestic metropolises for making their regional physical plans and giving permission to the construction of a landfill is affected by the level of political acceptance. The final decisions about landfill location will be disrupted by the political unwillingness.

2. Acceptance by the pressure groups local?

In the decision-making process, an important factor is the acceptance of landfill by the public in their region or metropolis. An attitude that is gradually becoming common is the so-called NIMBY (not in my backyard) syndrome. In the presence of well-organized local groups who have good relations with the local authorities and the media (papers, radio, and television), the impact of the public is significant. The distance the local pressure groups are successful in the prohibition of the decision-making process indicates the level of public acceptance.

3. Property of the landfill area

A very important issue is the ownership of the needed land for the landfill. Due to additional problems with the cost of the land in private ownership, public ownership is considered easier. Occasionally, confiscation is required and it leads to delays.

2.10.2.3   Financial and Economic Criteria

1. Land costs

Considerably different land prices for each location determine the costs of land. The degree of compromise for the owner or actual users is influenced by the price for which the actual use of the land is important. The most preferable potential landfill is the one with the lowest costs.

2. Costs for the access of the landfill

The availability of closed roads to the landfill influences the costs to access it. The costs will increase, in the case of the necessity of reconstruction of actual roads. However, location of a landfill site is significantly affected by the presence of the road network.

3. Transport costs

Transport distances from the source of waste generation, means of transportation and the form of collection, are important factors in transport costs. The demand for waste transfer stations and the feasibility to use railways are other affecting factors of transport costs.

4.   Costs for personnel, maintenance and environmental protection

The various potential landfill sites will not differ considerably in costs for personnel. The accessibility of the needed soil for the daily of regular covering and the stability of the landfill are determining factors of maintenance. Alteration of the maintenance costs will be the result of a lack of soil in the area and the consequent importing. The pollution of the soil, groundwater and surface water at the landfill should be prevented through the placement of additional technical supplies. Furthermore, observing the drainage system and the quality of the leachate and surface water are significant factors in the maintenance costs. The more suitable potential landfill is the one with the lowest maintenance costs.

5.   Costs for the after-care

Provisions to monitor the groundwater quality, the existence of gas, the winning of gas and stability of the completed landfill also influence the costs for after-care in addition to the kind of final use. Required supplies are attached to the characteristics of the filled waste, the kind of subsoil, the hydro-geological situation and the kind of final use.

2.10.2.4   Hydrologic/Hydrogeological Criteria

1.   Surface water

Surface water will be protected from contamination by leachate if the landfill site is not placed within surface water or water resources protection areas. In order to prevent waste from entering rivers and important streams, it should take safe distance from the meandering and non-meandering rivers. A distance of 100 feet (30.48 m) of any non-meandering stream or river, and at least 300 feet (91.44 m) from any meandering stream or river, should be taken into account for locating a landfill. The aquatic habitats can be secured from blown debris and runoff from the large ponds, lakes, and reservoirs through a buffer zone of land. A minimum distance of 100 feet (30.48 m) from any landfill site should be considered for large bodies of water (greater than 20 acres (80,937.45 m^2) of surface area). Exclusion of whole watershed draining into the reservoir from landfill sites is needed if surface water impoundments supply the regional drinking water. The contamination will be finally diluted if the velocity of the surface flow is high. The highest-ranking score will be attributed to the potential landfill location of the highest velocity of the overland flow. The downstream effect from carried away waste in the case of higher water levels contributes to the major worries about sitting landfills within floodplains. Any landfill shouldn't be placed in the area of the floodplains of major rivers due to higher discharge and huge downstream impact of major rivers. It is not secure to construct

Table 2.5 Groundwater
depth and landfill suitability

Depth to groundwater	Suitability
Over 60 m	High
15–60 m	Moderate
Under 15 m	Low

Table 2.6 Groundwater
quality and landfill suitability

Groundwater quality (TDS in mg/l)	Suitability
Over 10,000	High
1000–10,000	Moderate
Under 1000	Low

a landfill within the 100-year flood stage of a minor river or stream (Pazoki et al. 2016; Şener et al. 2006).

2. **Groundwater**

Landfills are not advised to be placed over high-quality groundwater resources, in order to protect subsurface drinking water. A compound liner system and monitoring wells should be used otherwise a fresh groundwater (total dissolved solids >1000 mg/l) should be avoided. Landfills should be located more than 304.8 m (1000 feet) up gradient from water wells, because of the travel down gradient of potential leachate leaks. Sites with aquifer depths -to-groundwater of 60.96 m (50–200 feet) are considered more appropriate than the ones with less than 15.24 m (50 feet) depths. Depth to groundwater and the number of dissolved solids based landfill suitability is presented in Tables 2.5 and 2.6 (Bagchi 2004; Bolton 1995).

The final spreading of leachate beneath the landfill is increased due to a high-speed rate of the groundwater flow. The porosity of the soil and the filtering speed are the main factors affecting the velocity of the groundwater flow. The more suitable location for a landfill is the one with the lowest velocity of the groundwater flow.

The risk of pollution of the groundwater or river water will be higher the ground-water level or a nearby river level is. The place with the lowest groundwater or river level is more preferable for the landfill location. The potential risk of polluting the groundwater is minimized when the layers in the subsoil are impermeable. Low permeability is especially related to clay layers. The more proper location for a landfill is the one where the subsoil layers have a high impermeability.

2.10.2.5 Topographical Criteria

Site selection, structural integrity, and the flow of fluids surrounding a landfill site are significantly influenced by the topography of an area, due to important effects for land capacity, drainage, land use, surface and groundwater pollution, roads to the site and it's operation ultimate. The topography of a site is straightly related to the decision of the type of landfill design (area-, trench-, and depression-type landfills).

The best sites for the area- and trench-type landfills are flat and smooth hills which are not likely to flooding. However, other land uses like agriculture, residential or commercial development prefer this kind of topography consequently the land prices increase. Contamination of groundwater sources of drinking water may be the result of depressions like sinkholes which are typically related to unstable caverns, so it should be avoided. Stone quarries, clay pits, and strip mines are other topographical depressions resulting from human activities. Low permeable formations such as clay, siltstone, or shale are included in the floor of these depressions. While the most appropriate material for depression-type landfills is clay pits, sand gravel pits should be discarded due to permeability except when the bottom formations are impermeable.

The degree or grade of the topography is connected to the potential for slope failure. Waste containment failure and release of debris into the surrounding area are the results of slope failure underneath or adjacent to landfills. Lands of greater slopes than 15% are not suitable for waste disposal sites. The flow of surface run-off, run-on, and drainage from a waste disposal site are directly influenced by the regional topography. The rainwater or leachate that drains overland away from the facility is called run-off while drainage overland on to any part of the facility is referred to as run-on. The desirable sites are the ones with the small requirements control of run-on from upland and slow run-off. Berms and stream diversion is used to control run-on. The speed of water crossing the site affects the run-off control. The portions of major drainage basins (watersheds) should not be selected as a landfill site in order to limit the possible spread of contaminated runoff.

2.10.2.6 Geological Criteria

The soil types made from the raw material, loading tolerable capacity of the landfill's foundation soil and the migration of leachate are straightly controlled by the geology of the area. The nature of soils and the permeability of the bedrock are specified by the rock and its structure type. The shift of leachate and potential rock-slope failure along joints and declined bedding planes are influenced by geologic structure.

Little (if any) will be transmitted by unfractured crystalline rocks while immediate transport of fluids is permitted in poorly cemented sandstones, as extreme permeability rates are compared. Sandstone is a less preferable landfill bedrock than other sedimentary rocks such as limestone and shale because of higher permeability rates. Susceptibility of the carbonate rocks to dissolution from low pH leachate which is related to discontinuities and karsts features such as collapses, sinkholes, and caverns makes limestone's more suitable than shales. The transmission of fluids in the bed is slowed or confined in shale's so that shale formations are well suited for landfill sites (Ghasemzadeh et al. 2017; Bagchi 2004; Bolton 1995) (Fig. 2.13).

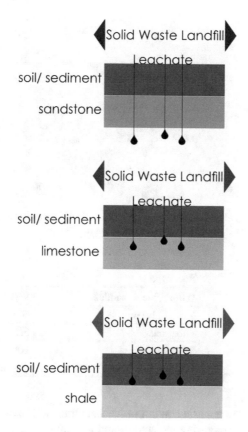

Fig. 2.13 Litho logic influence on permeability rates of landfill leachate

Figure 2.14 shows the influence of sedimentary rock types on the permeability rates of landfill leachate. Table 2.7 summarizes some of the various rock types of suitability for landfill siting.

The movement of leachate and on the structural integrity of the bedrock substance is directly affected by the structure and orientation of discontinuity planes. The possibility of rock-slope failure along discontinuities exists for sites made of inclined rocks greater than 45-degree dip and therefore is an unstable site. Down-dip directions will be followed by the leachate flow. The axis of anticlines and structural domes should not be chosen for landfill sites in order to confine the spread of leachate. Anticlines and domes are not only caused by the spreading of landfill leachate but also are usually related to oil and natural gas fields and should be avoided. In contrary to anticlines and domes, the best sites for leachate to pool into are synclines and structural basins. The impact of geologic structure on the dispersing or collection of landfill leachate is illustrated in Fig. 2.14 (Schwartz 2001).

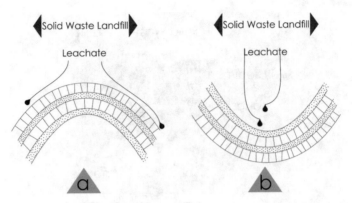

Fig. 2.14 Influence of geologic structure on the transport of landfill leachate **a** spreading of leachate over anticline axis, whereas **b** the pooling of leachate within syncline

Table 2.7 Landfill suitability of bedrock

Rock type	Suitability
Unfractured crystalline	Very high
Shale and clay	High
Limestone	Fair to poor
Sandstone	Poor to very poor
Unconsolidated sand/gravel	Unsuitable

As a fault is able to act as a conduit of leachate transport reducing the structural integrity of bedrock supporting the landfill and its equipment, faulted areas are not appropriate for landfill.

2.10.2.7   Availability of Construction Material

A satisfactory source of soil with suitable textures for daily and final covering is involved in sanitary landfill design.

Three basic reasons have consisted of the significance of soil in landfill development:

– Cover: after completion of the landfill area, the cover is the material applied in covering of the solid waste daily. The quantity of the generated leachate is considerably affected by the permeability of the final cover.
– Migration control: Leachate and methane movement from the landfill are aren-trolled by the material. The movement will be deaccelerated through an imperme-able formation; the Installation of more controls in the landfill is required when less protection is lessened in permeable soil.
– Support: A proper soil for the construction of the landfill is the one beneath it. It is a rigid foundation for liners, roads, and other construction.

Table 2.8 Soil textures and landfill suitability

Soil type	Suitability
Silt to very fine silty clay	Very high
Clay	High
Mixed	Moderate
Sandy	Low
Clean sand/gravel	Unsuitable

A fairly low permeability, sufficient bearing capacity to support facilities, a pH of at least 5, a low permeability potential and a high attention exchange capacity, are optimal features of landfill soil. However, the features of a site's soil like depth may vary and the soil volume may not be enough. Finely grained soils are more preferable for landfills than crude grained soils, as it is stated in Table 2.8. The suitability of a clay soil compared to soils with a silty clay texture is often reduced due to clays characteristics such as low drainage rates, shrink/swell potential, and low workability.

Corruption of the bottom liners and/or the drainage system is the result of the unstable foundation of the landfill caused by a great sensitivity to soil consolidation (peat and clay soils). A more suitable place for a landfill is the location of the lowest susceptibility to soil consolidation.

2.10.2.8   Other Criteria

1. Residential and urban areas

According to regulation on solid waste control in Turkey, it is not permitted to construct landfills on sites within a distance of fewer than 1000 m to settlements. The Ministry of Environment and the highest local authority and the concerned municipality will allow the construction of landfills in a distance less than 1000 m to settlements, only in the presence of natural barriers like hills, trees or forests between the landfill site and the settlements.

2. Military areas

The testing area of military equipment or training of military personnel is not accessible for public usage.

3. Airports

Birds are especially attracted by organic waste and due to potential danger of birds for airplanes; landfills should be placed at a specific distance from airports (3048 m).

4. Industrial areas

Potential landfill sites may include industrial areas. Any industrial area or its vicinity may be a suitable place for a landfill, but it is highly dependent on the kind of industry.

For example dust or food factories may be excluded. The infrastructural provisions prepared in an industrial area, are considered as an advantage.

5. Difficult infrastructural provisions

Making the suitable location for the application as a landfill will be very problematic in the presence of infrastructural provisions such as cables, roads or existing plans for drainage. For example, no landfill should be built within 300 m of the highway.

6. Climate

Due to connection of climate features such as prevailing winds, precipitation, evapo-transpiration and temperature variations to odors, dust, leachate generation, blowing litter, cover soil and erosion, climate characteristics should be considered in the site selection process.

2.11 Design of Landfills

The following are the significant considerable topics in the design of a landfill (though not necessarily in the order given) (Tchobanoglous and Theissen 2005; Bagchi 2004):

1. Layout of landfill site
2. Types of wastes that must be managed
3. The requirement for a convenient transfer station
4. Estimation of landfill capacity
5. Geology and hydrogeology of the site
6. Selection of leachate management facilities
7. Selection of landfill cover
8. Selection of landfill gas control facilities
9. Surface water management
10. Aesthetic design considerations
11. Development of landfill operation plan
12. Determination of equipment requirements
13. Environmental monitoring
14. Public participation
15. Closure and post-closure care.

The following sections consider the development of an operational plan for a land-fill and the determination of equipment requirements. Closure and post-closure care deals with environmental monitoring. Table 2.9 has described the important factors to be considered in the design of landfills. The final use or uses of the completed site should be carefully taken into account during the development of the engineering design report. Only dirt should fill the land reserved for administrative offices, build-ings, and parking lots. The underground gas should be prevented from migration to

	Parameter	Range	Median
Table 2.9 Design guidance for leachate-collection system components	Leachate loading rate (gpd/ac)	600–1000	750
	Maximum leachate head (in.)	9–12	11
	Pipe spacing (ft)	60–400	180
	Collection pipe dia. (in.)	6–8	8
	Collection pipe material	PVC or HDPE	HDPE
	Pipe slope (%)	0.5–2	1
	Drainage slope (%)	0.2–2	1

buildings through sealing. Membrane seals or soil gas extraction systems can be used for protection.

2.12   Landfilling Methods

The possible classification of the principal methods used for landfilling of MSW is (1) excavated cell/trench, (2) area and (3) canyon. Figure 2.15 illustrates the principal features of these types of landfills which are described as follows.

2.12.1   Excavated Cell/Trench Method

If a sufficient depth of cover material is available at the site and the water table is not near the surface, the ideal method of landfilling is the cell/trench method (see Fig. 2.15a). Cells or trenches excavated in the soil are typically used to landfilling of solid wastes (see Fig. 2.15a). Daily and final cover will be the excavated soil from the site. The movement of both landfill gases and leachate is limited by the lining of the excavated cells or trenches with synthetic membrane liners, low permeability clay, or a combination of the two. The typical shape of excavated cells is square, up to 1000 ft in width and length, with side slopes of 2:1 to 3:1. The length of trenches differs from 200 to 1000 ft, the depth from 3 to 10 ft and widths from 15 to 50 ft. The artesian or region of a saturation landfill is a variation of this method. The landfills are built beneath the naturally occurring groundwater table surface. The passage of groundwater into the landfill cell is controlled by drainage systems. According to this method lined and unlined sites have been founded both.

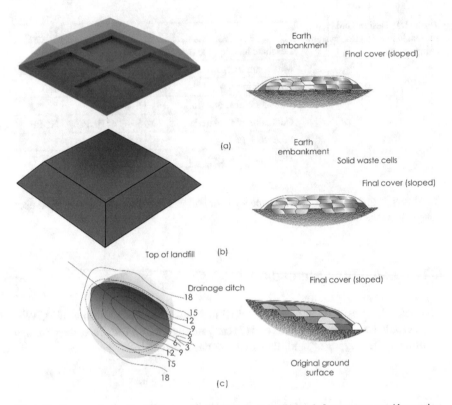

Fig. 2.15 Commonly used landfilling methods: **a** excavated cell/trench; **b** area; **c** canyon/depression

2.12.2 Area Method

When the terrain is inappropriate for the cell excavation or the placement of the solid wastes in trenches, the area method is employed (see Fig. 2.15b). The application of area-type landfills is vital in places such as many parts of Florida and elsewhere, where the groundwater conditions are high. The foundation of a liner and leachate management system is a constituent of the site preparation process. Harvesting of cover material must be performed by truck or earthmoving equipment from adjacent land or from borrow-pit areas. The intermediate cover material can successfully include compost produced from yard wastes, MSW, foundry sand, and auto shredder fluff in placed where the suitable material for cover is limited. Application of movable provisional cover materials such as soil and geosynthetics is an alternative technique. Prior to the initiation of the next lift, Soil, and geosynthetic blankets, placed briefly over a completed cell can be eliminated.

2.12.3 Canyon/Depression Method

Canyons, ravines, dry borrow pits, and quarries have been used for landfills (see Fig. 2.15c). The geometry of the site, the features of the accessible cover material, the hydrology and geology of the site, the type of leachate and used gas control facilities and the access to the site determine the techniques of placement and compaction of solid wastes in canyon/depression landfills. An important factor in the progression of canyon/depression sites is the management of surface drainage. The aggregation of water behind the landfill is prohibited typically through the start of filling at the head end of the canyon and ending at the mouth. The Canyon/depression sites are similarly operated as the methods previously described as they are filled in multiple lifts. The excavated cell/trench method discussed previously, may be applied to perform the initial landfilling in the case of flatness of a canyon floor.

2.13 Other Types of Landfills

Meeting certain objectives is made possible with the construction of other different configurations of landfills. The construction and destruction of waste landfills which only accept resulting materials of the destruction of buildings and removing roadways are an example. Landfills related to acceptance of high volumes of industrial waste such as that from paper mills, foundries, power plants, and mines are another type of specialized landfill. Particular design factors are considered for each of these landfills. According to the unique specialized nature of the accepted waste, any of the common design elements may not be included in the design of the landfill. For instance, due to the pre-elimination of all the organic matter during the combustion process and consequently no decomposition of waste is expected so there is no requirement of a gas recovery system for a power plant ash landfill. A bioreactor stabilizes waste more quickly so that emerging technology is applied. The manner a bioreactor landfill is constructed and operated, improves the rate of decomposition of the organic material within the municipal solid waste. Rapid initiation of the decomposition of the waste takes place due to adjusted operating procedures as compared to conventional landfills. Recovery of the methane gas occurs due to the gas collection facilities which are installed promptly upon the construction of the landfill cell. Recycling of the withdrawn leachate from the base of the landfill and addition of other sources of moisture, such as sewage sludge to the waste profile, leads to the acceleration of the decomposition rate. The reduction of the long-term observation period of landfills after their closure is an option of bioreactor landfills due to the rapid stabilization of the waste. Furthermore, improve in the amount of disposable waste on the original landfill site as a result of reducing the waste volume to the maximum extent possible and in the shortest period of time, is a goal of some designs. The methods for lining and covering bioreactor landfills are still under consideration. The factors related to slope stability, landfill liner leakage, methods for collecting landfill gas in an almost opened cell and minimization of odor by the construction of leachate recirculation systems are the design issues that are being assessed nowadays.

2.14 Liner Materials

Migration of leachate from the landfill and to facilitate removal of leachate is prevented by the liner system. Natural material and/or geomembranes are selected for their low permeability in the liner system. Natural clays or clayey soils are usually used to construct soil liners. Commercial clays (bentonite) can be mixed with sands to produce suitable liner material in the case of a lack of readily available natural clay materials. Geomembranes are impermeable thin sheets composed of synthetic materials, like polymers. Due to its resistance to most chemicals found in landfill leachates, high-density polyethylene (HDPE) is likely to be used in MSW landfill liners most commonly.

Depending on the in-use local, state or federal regulations, landfills may be designed with single, composite or double liners (see Fig. 2.16). Clay or a geomembrane is used in the building of a single liner. The minimum needed liner by RCRA Subtitle D, called a composite liner, has two layers (Vesilind et al. 2002b): A clay material is in the bottom and a geomembrane in the top layer. In order to minimize leakage, these two layers are in confidential contact. Either two single liners or two composite liners may be constituted in a double liner (or even one of each). A synthetic liner on a side slope, prepared for the earth cover is depicted in Fig. 2.17 (Vesilind et al. 2002b). Beneath this synthetic liner, a clay layer has been already installed.

A leachate collection system is included in each liner. A series of pipes placed between the liners to collect and observe any leachate leaking through the top liner is the leak detection system is the collection system separating the two liners. The top component in the double liner system is composed of the geosynthetic clay liner, recently. A thin clay layer (usually sodium bentonite) is introduced in

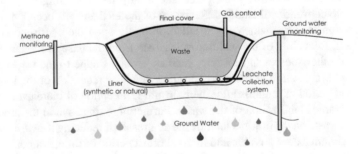

Fig. 2.16 Design components in a subtitle "D" landfill

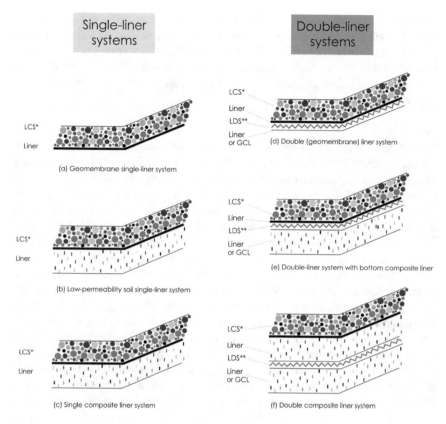

Fig. 2.17 Examples of liner systems in municipal solid waste landfills. LCS—leachate collection system, GCL—geosynthetic clay liner, and LDS—leachate-detection system

this liner, supported by geotextiles (a geosynthetic filter) or geomembranes. More volume is permitted to be used for waste deposition, due to less volume consumption by the geosynthetic clay liner which is placed easily in the field. Obviously, the degree of protection of the liner system is proportional to the number of layers included. However, the expenses will dramatically increase. The cost of a composite liner maybe like $250,000 per acre. An exhaustive quality control/quality assurance program is necessary during liner installation due to criticality of the liner in groundwater protection.

2.14.1 Soil and Clay Liners

Soil liners are applied in single liner systems and also in composite liner systems. A soil liner may be either the single liner system or composite liner system. Movement of the leachate from the fill into the subsurface environment is reduced or even blocked when a soil liner is used as a single liner. A soli liner provides protective bedding for the overlying soft membrane liner (FML) and it serves as a backup for breaches in the FML as the lower component of a composite liner. All soil liners act as a long-lasting stable structure as a base for overlying works and materials, which is a useful function.

Materials
Permeability of the soil must be low enough (less than $I \times 10^{-7}$ cm/s) after compaction under field conditions to be adequate as a liner. The liner should be strong enough to support it and the overlying materials after being compacted. Construction equipment is used to handle the liner material which should yield for the process. Ultimately, when a liner material is exposed to waste or leachate from the waste, its permeability should not significantly reduce. Mixing a soil deprived in specific features with another soil or with a soil additive may make it suitable. The addition of bentonite cement in order to decrease permeability is an example. Necessary information about the interrelationship between moisture content, density, compaction effort, and permeability is provided by determination of the compaction and permeability characteristics of the selected soil liner material through laboratory tests. Well-compacted clay soil is the most commonly employed available material. A clay liner usually is made as a membrane up to 1 m thick. The clay membrane must be continually moist in order to work as a liner. Natural clay additives (e.g., montmorillonite) may be disked into it to form an effective liner if enough clay is not available locally. Optimum types and amounts of additives should be determined. The native soil at the site would be the best in cost and convenience if required specifications are met. Otherwise, suitable soil must be imported. In the case of the application of off-site material, transport cost becomes an obviously important consideration. An established borrow pile stored at the site is composed of the excavated locally or imported liner material. The soil should contain at least 20% fines, the plasticity index (PI) must be 10–30%, crude fragments should be monitored to no more than about 10% of gravel-size particles, the material should not contain rocks larger than 2.5–5 cm in diameter (Vesilind et al. 2002b; Delarestaghi et al. 2018; Rushbrook and Pugh 1999) (Fig. 2.18).

2.14.2 Flexible Membrane Liners

Premade polymeric sheeting is the constituent material of a flexible membrane liner (FML). A flexible liner may be applied in many ways. It may be used as a single liner installed directly over the foundation soil, for instance. It may be a fraction of a composite liner placed upon a soil liner, on the other hand. Ultimately, it may be housed above or below a leak detection system in a double-lined landfill.

The selection of the FML material, designing the subgrade and planning the installation are the major steps included in the application of a flexible membrane liner. The

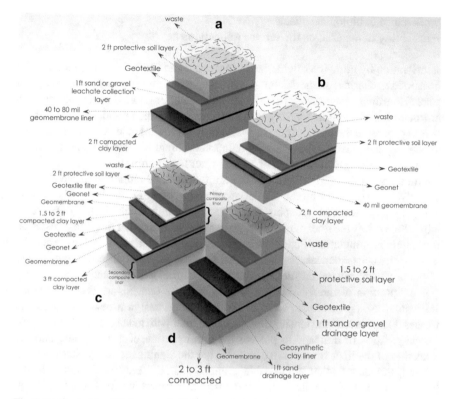

Fig. 2.18 Typical landfill liners: **a, b** single composite barrier types; **c, d** double composite barrier types

arrangement of subcomponents, such as sealing and anchoring systems and vents is the last step. high-density polyethylene (HDPE) chlorinated polyethylene, chlorosulphonated polyethylene, and polyvinyl chloride are some of the membranes type commonly used for lining sanitary landfills. Selection of a FML follows important criteria like:

- chemical compatibility with the contained leachate
- possession of appropriate physical properties like thickness, flexibility, strength, and degree of elongation
- availability and cost.

Listed specifications by the manufacturer are the basis to judge the compatibility, in the absence of testing facilities. Other considerable mechanical properties involve:

- stiffness or flexibility at different temperatures, resistance to puncture
- thermal expansion
- seaming characteristics vs. resistance to weathering

– resistance to biological attack
– instability of material on the service impoundment.

Deterioration by ultraviolet light, ozone reactions, and plasticizer migration are the forms of weathering. Bacteria, fungi, and rodents are the vectors of a biological attack. Here again, the provided data by the manufacturer is relied on. Obtaining such information is difficult in contrary to the availability of some published literature.

A key factor in the maintenance of FML purity is the subgrade upon which it rests. Supporting and prevention of the accumulation of liquid beneath the liner is a function done accordingly. Infiltration of groundwater from surrounding soils may cause accumulation of liquid. Furthermore, the uplift stress and reduction of the strength of underlying soils may be the results of accumulation. Surrounding soils may be polluted by the leachate escaping from the fill through breaks in the membrane. Subsidence below the liner may lead to mechanical stresses other than those resulting from liquid accumulation. Punctures and tears, and repeated stresses that abrade the liner are the consequences of concentrated stresses, differential movements of the subgrade causing tangential stresses which are the form of other mechanical stresses.

Configuration of the subgrade to be free of crude changes in grade and plane and regular as is possible are some of the foundation design measures. Tangential stresses do not exceed the tensile strength of the liner with managed sidewall slopes. The preparation of drainage to prevent the accumulation of gas or liquid and the protection of the liner from being punctured are the significant design factors.

Sand, gravel, or other comparable granular material may be included in the drainage layer. A geotextile (a fabric devised to provide tensile strength and act as a filter protected by a layer of lower permeability soils) is a possible alternative form. The following are the related problems with drainage layers:

– difficult to install on slopes not stable on sharp slopes
– likely to eruption by workers during construction
– can be disrupted by wind or water during construction
– risk of the liner being punctured by damaged or displaced pipes.

Elimination of rocks (larger than 25 mm), roots, and other debris from the surface should be included in surface preparation. Settlement and gas production under the liner will be minimized by removing organic material. In the end, a hard and unyielding base for the liner is provided if the substrate soil surface is compacted (Fig. 2.19).

The real installation of a flexible membrane liner is a complex task. It should be carried out by a qualified and competent company under the supervision of the manufacturer or one designated by the manufacturer and should be independently quality assured.

Fig. 2.19 Installation of a flexible membrane liner

2.15 Leachate and Stormwater Management, Collection, and Treatment

Efficient drainage layer constituting sand, gravel or a geosynthetic material is used to direct the leachate to low points at the bottom of the landfill. Perforated pipes are housed at low points to gather leachate and are sloped to permit the moisture to move out of the landfill.

Minimization of the potential for groundwater pollution is the initial purpose of lining a landfill cell. A barrier between the deposited waste and the groundwater and is provided by the liner and it is a basin for leachate produced by the landfill. The leachate produced by the landfill is restricted according to the RCRA Subtitle D regulations, so the collected leachate within the cell must be eliminated from above the liner as quickly as possible. Typically there are two methods for removal of leachate: gravity flow or pumping. A leachate collection system for an MSW landfill typically includes the following various components:

– Protective and drainage layers
– Perforated collection lateral and header pipes
– Pump station sump
– Leachate pumps
– Pump controls
– Pump station appurtenances
– Force main or gravity sewer line.

General guidelines for leachate collection system components based on a survey of landfill design engineers are provided in Table 2.9 Prior to the time the removed leachate from the landfill cell(s) can be treated, recirculated or transported off-site for final treatment and disposal, it is briefly stored on site. Equalization of flow quantities and basic quality to protect downstream treatment facilities is considerably affected by the storage of leachate. Surface impoundments and tanks are the typical leachate storage alternatives. Leachate Collection System Design Equations and Techniques due to federal regulations[24] that limit leachate, head to 12 in. (30 cm) on top of the liner; Prediction of this value has been devoted much attention. This depth on the liner is controlled by the drainage length, drainage slope, permeability of the drainage materials, and the leachate impingement velocity. Development of an equation that predicts the leachate depth on the liner can be based on Darcy's law (in conjunction with the law of continuity) which is founded on anticipated infiltration rates, drainage material permeability, distance from the drain pipe and slope of the collection system.

For $R < 1/4$,

$$Y_{max} = \left(R - RS + R^2 S^2\right)^{1/2} \left[\frac{(1 - A - 2R)(1 + A - 2RS)}{(1 + A - 2R)(1 - A - 2RS)}\right]^{1/2.4}$$

For $R = 1/4$,

$$Y_{max} = \left(R - RS + R^2 S^2\right)^{1/2} exp\left[\frac{1}{B}\tan^{-1}\left(\frac{2RS - 1}{B}\right) - \frac{1}{B}\tan^{-1}\left(\frac{2RS - 1}{B}\right)\right]$$

For $R > 1/4$

$$Y_{max} = \frac{R(1 - 2RS)}{1 - 2R} exp\left[\frac{2R(S - 1)}{(1 - 2RS)(1 - 2RS)}\right]$$

where:
   $R = q/(K \sin^2)$, unitless.
   $A = (1 - R)^2$, unitless.
   $B = (4R - 1)^2$, unitless.
   $S = \tan$, the slope of liner, unitless.
   Y_{max} = maximum head on liner, ft.
   L = horizontal drainage distance, ft.
   α = inclination of the liner from horizontal, degrees.
   q = vertical inflow (infiltration) per unit of horizontal area, ft/day.
   K = hydraulic conductivity of the drainage layer, ft/day.
   These equations are clearly difficult to use and a more conservative but far-easier-to-use equation has been proposed:

$$Y_{max} = \frac{P}{2}\left(\frac{q}{K}\right)\left[\frac{K\tan^2\alpha}{q} + 1 - \frac{K\tan\alpha}{q}(\tan^2\alpha + \frac{q}{K})^{1/2}\right]$$

where:

Y_{max} = maximum saturated depth over the liner, ft.
P = distance between collection pipes, ft.
q = vertical inflow (infiltration), defined in this equation as from a.
25-year, 24-h storm, ft/day.

This equation can be employed to calculate the maximum permitted pipe spacing based on the maximum allowable design head, anticipated leachate impingement rate, slope of the liner, and permeability of the drainage materials. If all other parameters are held constant, the height of the height is proportional to the distance of pipes (at greater construction cost), as suggested by the last equation. A lower hydraulic driving force through the liner and likewise reduce in the consequence of a puncture in the liner is the result of a reduced head on the liner.

Example: Determine the spacing between pipes in a leachate collection system using granular drainage material and the following properties.

Assume that in the most conservative design all stormwater from a 25-year, 24-h storm enters the leachate collection system.
Design storm (25 years, 24 h) = 8.2 in = 0.00024 cm/s.
Hydraulic conductivity = 10–2 cm/s.
Drainage slope = 2%
Maximum design depth on liner = 15.2 cm.

Solution:

$$P = \frac{2Y_{max}}{\left(\frac{q}{K}\right)\left[\frac{K\tan^2\alpha}{q} + 1 - \frac{k\tan\alpha}{q}\left(\tan^2\alpha + \frac{q}{K}\right)^{1/2}\right]}$$

$$P = \frac{2(15.2)}{\left(\frac{0.00024}{0.01}\right)\left[\frac{0.01(0.02)^2}{0.00024} + 1 - \frac{0.01(0.02)}{0.00024}\left((0.02)^2 + \frac{0.00024}{0.01}\right)^{1/2}\right]} = 1428\,\text{cm}$$

The efficiency of the leachate collection systems has been increased over such natural materials as sand or gravel using geosynthetic drainage materials (geonets). Highly efficient in-plane flow capacity (or transmissivity) is provided by a geonet which is composed of a layer of ribs overlaid on each other. The spacing between the collection pipes can be calculated from the following equation, in the case of application of geonet between the liner and drainage gravel:

$$\Theta = \frac{qP^2}{4Y_{max} + 2P\sin\alpha}$$

where Θ = transimissivity of geonet, ft^2/day

2.15.1 *Leachate Collection Facilities*

The leachate accumulating downhill a landfill is usually collected by the application of a series of sloped terraces and a system of collection pipes. The accumulated leachate on the surface of the terraces will drain to leachate collection channels as a result of the slope of the terraces as presented in Fig. 2.20a. The collected leachate is transferred to a central location, from which it is removed for treatment or reapplication to the surface of the landfill using a perforated pipe which is located in each leachate collection channel (see Fig. 2.20b) (Ghasemzadeh et al. 2017). The cross slope of the terraces is typically 1–5%, and the slope of the drainage channels

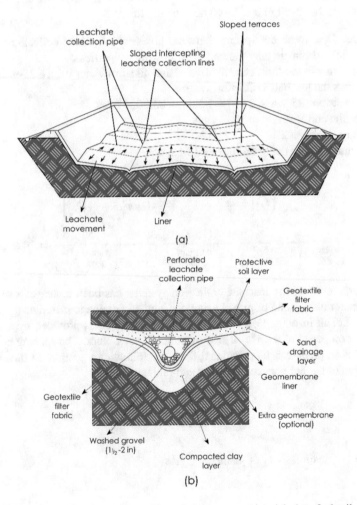

Fig. 2.20 Leachate collection system with graded terraces: **a** pictorial view; **b** detail of typical leachate collection pipe

is 0.5–1.0%. The equations devised by Wang are used to analyze the configuration and slope of the drainage system. The depth of leachate above the liner is determined by the cross slope and flow length of the terraces. The higher head build-up will be the result of smoother and longer slopes. Prevention of the leachate to pond at the bottom of the landfill in order to make a considerable hydraulic head on the landfill liner (less than 1 ft at the highest point, as specified in the federal Subtitle D landfill regulations) is the goal of the arrangement. The depth of flow in the perforated drainage pipe increases gradually from the upper reaches of the drainage channel to the lower reaches. The drainage channels are connected to a larger cross-collection system in huge landfills (Tchobanoglous and Theissen 2005; Pazoki et al. 2017).

2.15.2 Leachate Removal and Holding Facilities

Removal of accumulated leachate within a landfill is performed based on two methods. The leachate collection pipe is passed through the side of the landfill in Fig. 2.21a. The pipe penetrating the landfill liner will be safe if great care applied using this method. Removal of leachate from landfills involves the use of a sloped collection pipe placed within the landfill through an alternative method (see Fig. 2.21b). When the recycling of the leachate or its treatment is required at a central location, leachate collection facilities are used. Figure 2.22 shows a usual leachate collection access basement (Tchobanoglous and Theissen 2005). A holding tank like the one depicted in Fig. 2.23 is used to remove the leachate from the landfill in some locations. The type of treatment facilities available and the maximum permitted discharge speed to the treatment facility will determine the capacity of the holding tank (Renou et al. 2008).

Regularly, 1–3 days' worth of leachate produced during the peak leachate production period is held in leachate-holding tanks. The safety is more in double-walled tanks so that these are preferred as the result of being more affordable as compared to a single-walled tank. To be corrosion resistant, careful selection of tank materials is critical.

2.15.2.1 Leachate Management

A key issue in the removal of the potential of polluting underground aquifers by landfills is the management of leachate if it forms. (1) Leachate recycling, (2) leachate evaporation, (3) treatment followed by spray disposal, (4) wetlands treatment and (5) discharge to municipal wastewater collection systems are some of the applied alternatives for management of the collected leachate from the landfill.

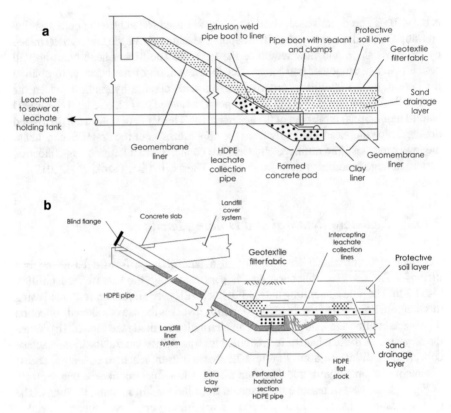

Fig. 2.21 Typical systems used to collect and remove leachate from landfills: **a** leachate collection pipe passed through the side of landfill; **b** inclined leachate collection pipe and pump located within the landfill

2.15.2.2 Stormwater Management

Diminishing of the leachate production is possible through many available operating and design controls like control of the size of the working face, placement of interim cover on the waste, and use of suitable stormwater runoff and run-on controls. Subtitle D of RCRA requires the control of stormwater run-on and runoff (Turner 1996). The production of leachate, erosion, and pollution of surface water is reduced as a result of the prevention of stormwater to enter the active area of the landfill due to run-on control. Minimizing run-on also confines the production of runoff from the landfill surface. Diversion of stormwater from active areas of the landfill can be prohibited by run-on. The ability to handle peak volumes generated by a 24-h to 25-year storm is required for any facility constructed to control run-on.

Contouring the land neighboring the landfill cell or constructing ditches, dikes, or culverts to divert flow are the components of the common measures to control run-on. Swales (see Fig. 2.25), ditches, berms, dikes, or culverts directing pollutant

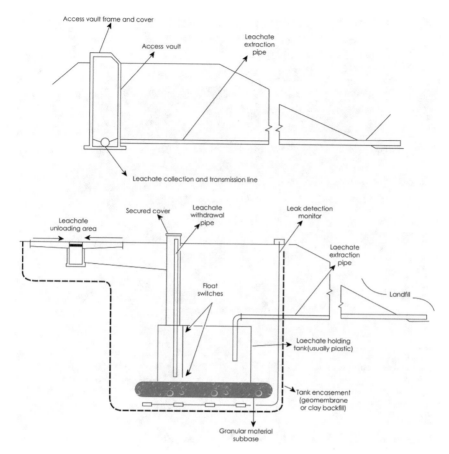

Fig. 2.22 A typical leachate collection access vault

runoff from active areas to storage and treatment facilities, and uncontaminated runoff from closed areas to detention facilities collect the generated runoff (Vesilind et al. 2002a). Collection and control of the volume generated from a 24-h to the 25-year storm of runoff from active areas are necessary. The arrangement of uncontaminated stormwater management facilities is commanded according to local regulations. For instance, in Florida, a quarantine pond must store the first inch of runoff for 14 days (Figs. 2.24, 2.25, 2.26, and 2.27).

The proper grading and appropriate design of the site drainage installed facilities are vital for those areas in which the stormwater runoff from the surrounding areas can enter the landfill (see Fig. 2.28).

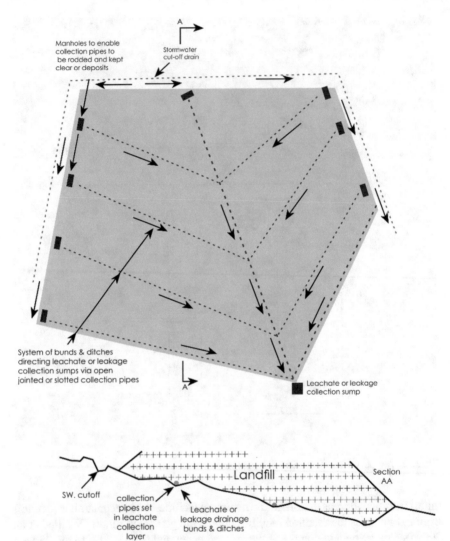

Fig. 2.23 Collected leachate in a holding tank

2.15.2.3 Storm Water Storage Basins

Installation of stormwater storage or retention basin may be required depending on the location and configuration of the landfill and the capacity of the natural drainage courses. In many cases, the construction of stormwater storage basins to hold the diverted stormwater flows in order to reduce downstream flooding is needed.

Normally, the collection of the stormwater should be done from the completed portions of the landfill as well as from areas yet to be filled. Figure 2.27 shows

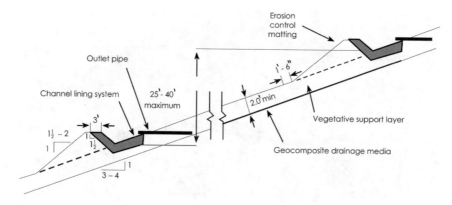

Fig. 2.24 Side slope swale in a landfill final cover

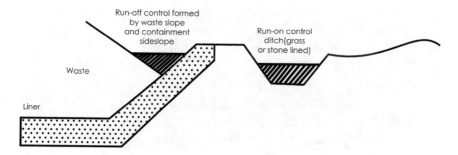

Fig. 2.25 Schematic of run-on and run off controls at a landfill

an example of a huge stormwater retention/storage basin (Vesilind et al. 2002a). To estimate the size of the stormwater sins, standard hydrological procedures are followed. Permission of discharge should be given by a federal or state authority about discharges from stormwater facilities.

2.16 Landfill Gas Management

One of the following three plans should be the basis of the gas management strategies:

– Controlled passive venting
– Uncontrolled release
– Controlled collection and treatment/reuse.

The advised emission for all MSW landfills is controlled passive venting. Allowance of uncontrolled release is just viable for small (less than 100 tons per day), shallow (less than 5 m deep) and remotely located landfills. If the gas concentrations are more

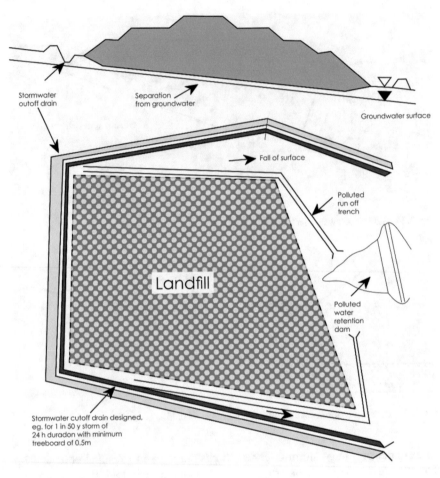

Fig. 2.26 Separating external surface water from a landfill

than acceptable limits, corrective measure (such as flaring) is done although landfill gas monitoring is carried out at all sites. When an experienced agency in the field proves the feasibility of a system, controlled collection and treatment/use will be established. Waste Management Practices: Municipal, Hazardous, Industrial (Barlaz et al. 1987).

2.16.1 Movement of Landfill Gases

The release of produced gases in soils to the atmosphere through molecular diffusion is a natural process under normal conditions. Both convection (pressure-driven) flow and diffusion will contribute to landfill gas release due to the usual greater than

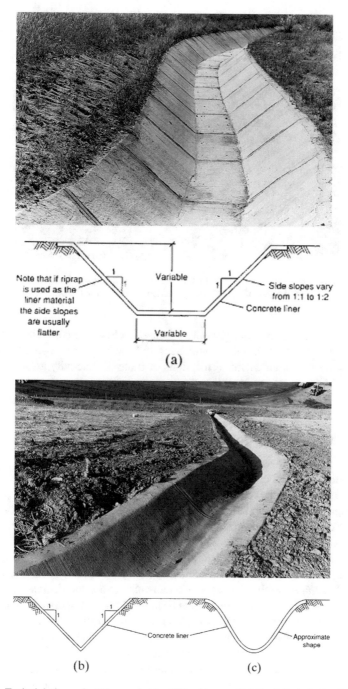

Fig. 2.27 Typical drainage facilities used at landfills: **a** trapezoidal lined ditch; **b** Vee lined ditch; **c** shaped-vee lined ditch. Note the trapezoidal ditch cross-section is expandable to accommodate a wide range of flows

Fig. 2.28 View of large stormwater retention/storage basin at a large landfill. The size of the basin can be estimated from the size of the vehicles parked at the bottom of the basin

atmospheric pressure internal pressure when a landfill is active. The sorption of the gases into a liquid or solid component and the release or consumption of a gas component through chemical reactions or biological activity are other factors affecting the movement of landfill gases.

2.16.2 *Movement of Principal Gases*

At a far distance of more than 400 ft from the borders of unlined landfills, both methane and carbon dioxide have been determined at concentrations up to 40% each, in spite of the greater tendency of the methane to escape to the atmosphere. At a distance of 1000 ft, methane concentrations over 5% have been measured. Several mechanisms are involved in the migration of landfill gas in unconsolidated soils. The gas is moved away from the landfill due to the combination of diffusion and the pressure of the gas in the landfill. The methane in the gas plume is oxidized through both aerobic and anaerobic processes which results in a faster decrease in methane concentrations than the expected amount accounted for by dispersion. Prediction of gas migration is additionally complicated due to geologic variations.

Accumulation of methane beneath buildings or closed spaces in the vicinity of a sanitary landfill (because its specific gravity is less than that of air) is the result of the allowed migration of methane to underground uncontrollably. No problem is caused by methane if proper venting is carried out except that is a greenhouse gas. Odor complaints and the required release controls are the consequences of odorous compounds and VOCs mixed with the methane.

– *Upward Migration of Landfill Gas*

Methane and carbon dioxide are the dominant gases released through the landfill cover into the atmosphere by convection and diffusion. An estimation of the diffusive flow through the cover can be provided by using:

$$N_A = D\alpha^{4/3}\frac{(C_{A_{atm}} - C_{A_{fill}})}{L}$$

where:
N_A = gas flux of compound A, g/cm^2·s (lb-mol/ft^2·d).
D = effective diffusion coefficient, cm^2/s (ft^2/d).
α = total porosity, cm^3/cm^3 (ft^3/ft^3).
$C_{A_{atm}}$ = concentration of compound A at the surface of the landfill cover, g/cm^3 (lb-mol/ft^3).
$C_{A_{fill}}$ = concentration of compound A at bottom of the landfill cover, g/cm^3 (lb-mol/ft^3).
L = depth of the landfill cover, cm (ft).
0.20 cm^2/s (18.6 ft^2/d) and 0.13 cm^2/s (14.1 ft^2/d) are the common values for the coefficient of diffusion for methane and carbon dioxide respectively. α gas = α if the dry soil conditions are assumed. The vapor flux from the landfill will be reduced accompanying reduction of the gas-filled porosity due to any infiltration of water into the landfill cover which is a condition with a safety factor if a dry soil condition is assumed. Normally, porosity values for different types of clay vary from 0.010 to 0.30.

– *Downward Migration of Landfill Gas*

Accumulation of the carbon dioxide will finally occur in the bottom of a landfill due to its density. Diffusive transfer of carbon dioxide from liner downward takes place through the liner and the underlying formation until it reaches the groundwater when a clay or soil material is used for liner (note that the movement of carbon dioxide can be limited with the use of a geomembrane liner). The hardness and mineral content of the groundwater will increase through solubilization as a result of the decrease in pH due to the high solubility of carbon dioxide in water.

2.17 Cover Requirements

Soil or similar inert material (such as ash from power stations, construction excavated materials, dewatered sewage sludge, dewatered river dredgings) should be employed throughout the lifetime of the landfill:

– to cover the landfilled wastes daily

- to provide intermediate cover protection to interim levels of fill, and, if appropriate,
- as the final cover on reaching design completion elevations.

2.17.1   Daily Cover

Soil (typically 6 inches) or alternative daily cover (such as textiles, geomembrane, carpet, foam, or other proprietary materials) is used to cover waste at the end of each working day. The goals of using a daily cover are the control of disease agents and rodents; reduction of odor, litter and air release; diminishing the fire hazard; minimization of leachate production. Easy maintenance and increase of slope stability are the results of the inclination of the landfill sides to a maximum slope of 1:3 is maintained. Before the upcoming deposition of waste on top of the location currently being filled, several weeks may be taken, commonly. There are many reasons about the use of daily cover:

- to reduce the attraction of wastes to birds and rodents and leave behind any food inaccessible
- to reduce the appropriate dwelling for flies and vermin
- to provide a better surface for vehicles traveling over the landfill
- to reduce exposure to atmospheric conditions by limiting air entering deposited wastes
- to reduce the scattering of light wastes (litter) by the wind
- to inhibit direct infiltration of rainwater into the waste
- Ideally, cover materials should be dug from within the site, thereby increasing its bare capacity. This will require a certain amount of stockpiling and double handling.

Sufficient space should be designated on the site for this goal. A landfill site can seldom arrange all of the necessary material for daily, intermediate and ultimate cover and any other required bund construction. Shortfall requirements will be imported, in such conditions.

2.17.2   Final Landfill Cover

The primary purposes of the final landfill cover are (1) diminishing the infiltration of water from rainfall and snowfall after the landfill has been completed, (2) confining the uncontrolled release of landfill gases, (3) suppression of the proliferation of vectors, (4) limitation of the potential for fires, (5) preparation of suitable surface for the revegetation of the site, and (6) serving as the central element in the reclamation of the site. The landfill cover must have the following characteristics to meet these goals:

- Be able to tolerate climatic extremes (e.g., hot/cold, wet/dry, and freeze/thaw cycles).
- Be able to withstand water and wind erosion.
- Have stability against slumping, cracking and slope failure, and downslope slippage or creep.
- Resist the effects of variations in landfill settlement caused by the release of landfill gas and the compression of the waste and the foundation soil.
- Resist failure due to surcharge loads emerging from the stockpiling of cover material and the travel of collection vehicles across completed portions of the landfill.
- Resist deformations caused by earthquakes.
- Tolerate variations to cover materials resulted by constituents in the landfill gas.
- Resist the disruptions caused by plants, burrowing animals, worms, and insects.

Regardless of the possibility of water incursion, all Subtitle D landfills have to be capped. Prevention of the leachate formation resulting in groundwater pollution is the main purpose. Landfills will be merely storage facilities due to maintained dry conditions preventing the process of biodegradation as a result of keeping water out of the landfill. The role of placement of the final cap after reaching the landfill its designed height is to diminish the infiltration of rainwater, to minimize dispersal of wastes, accommodate settling, and facilitate long-term maintenance of the landfill. Vegetation and supporting soil, a filler and drainage layer, a hydraulic barrier, foundation for the hydraulic barrier, and a gas-control layer are the constituting elements of the cap (from top to bottom). A schematic representation of a recommended top slope cap is depicted in Fig. 2.29 (Vesilind et al. 2002a). A top cap should be less permeable than the bottom liner according to EPA regulations.

Critical considerations for landfill caps are slope stability and soil erosion. The buildup of water and/or gas pressures results in the decrease in the contact stresses between layers and drainage forces should be withstood by the interface friction between adjacent layers with the common side slopes of 1:3 to 1:4. So, composite liner caps (geomembrane placed directly on top of a low permeability soil) are not advised on slide slopes.

2.18 Landfill Construction

The landfill is an area of construction long associated with The Pryor Group. Extremely high specifications should be adopted in a landfill site construction to meet all the devised standards by the Environment Agency and other authorities. The geology of the area, the water table, and any adjacent rivers or other watercourses are the significant factors in the building of cells which differ from site to site. A landfill site should be suitable for use in at least next 25 years due to investment in infrastructure and environmental protection measures. Other important planning factors are:

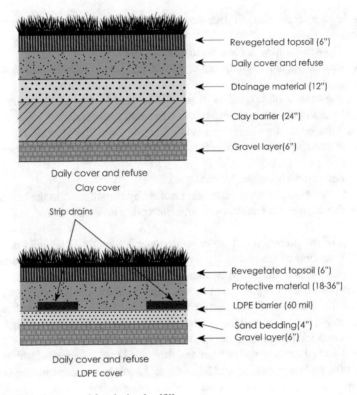

Fig. 2.29 Typical caps used for closing landfills

- Roads for the waste collection vehicles
- Type of waste
- Volume of waste
- Other related operations such as composting areas, sorting areas, etc. (Fig. 2.30).

Fig. 2.30 Landfill construction

2.18.1 Construction Equipment

Possession of the appropriate equipment to perform the job is necessary for the contractor in the construction of a compacted clay liner. The absolute minimum of the following equipment is considered necessary by the author:

- One (1) CAT 815 compactor, or equivalent;
- One (1) track-mounted excavator;
- One (1) bulldozer (D-7, minimum);
- One (1) harrows disc or roan plow;
- Three (3) dump trucks;
- One (1) 10,000-gallon water truck, with water canon;
- One (1) motor grader;
- One (1) smooth drum roller;
- Diesel fuel storage tank enough to maintain the equipment (estimate an average of 10 gallons per hour, per machine);
- Pumps (typically 6-inch minimum) and hose sufficient to dewater the cell and borrow area(s) (Fig. 2.31).

The above list is an absolute minimum; but it is needed to reach a production speed of about 1, 250 cubic yards per day. A contractor suggesting to carry out the work with less equipment is strictly doubted in experience and ability. Clearly, a potential risk threatens the work done with the minimum amount of equipment with no backup. It is possible for the contractor to operate with one of his dump trucks; although if his excavator ruptures, the operation will be ceased. As a result, in the case of suggestion to do the operation with only the minimum number of necessary pieces, the age of the equipment is a significant factor. In opposite, if more dump

Fig. 2.31 Landfill compactor used to compact waste placed in a landfill

trucks are available, but only a single compactor or excavator is present to load the trucks, the production rate will be the same (Bagchi 2004; Tang et al. 1994).

2.18.2 Selection of Landfill Cover

As discussed earlier, several layers performing a specific function are usually comprised of a landfill cover. Confining the entrance of surface water and controlling the release of landfill gases are the consequences of the common application of a geomembrane liner as a barrier layer. The place of the landfill and the local climatological conditions are important factors in the selection of a certain cover configuration. The application of a profound layer of soil is preferred by some designers for example. The minimum slope of almost 3 to 5% should exist in the final cover, assuring the fast elimination of rainfall from the completed landfill and avoiding the formation of puddles. In the case of slopes greater than 25%, cover slope stability must be considered during design.

2.18.3 Selection of Gas Control Facilities

Most new landfills have gas collection and treatment facilities to control the release of landfill gas, notably methane leading to the greenhouse effect and its possible migration underground to potentially cause explosions or kill vegetation and trees. The amount of landfill gas must be primarily estimated employing the outlined methods in order to determine the size of the gas collection and processing facilities.

Analysis of several rates is necessary due to variations in rate gas production rates which depend on the operating procedures (e.g., without or with leachate recycle). Determination of the rate of gas production with time is the next step. The arrangement and capacity of the landfill determine the decision to use horizontal or vertical gas recovery wells. The capacity of the landfill site and the chance to sell the produced power during the conversion of landfill gas to electrical energy or the access to the utility boilers for energy recovery affect the decision to burn or recover energy released from the landfill gas. Gas collection equipment is not employed regularly in many small far located landfills.

2.18.4 Materials

Materials used in landfill construction must be expected to withstand a wide range of natural stresses for very long periods. Many materials deteriorate over time when exposed to chemicals existing in leachate.

Fig. 2.32 Application of geomembrane in landfill

Landfill owners and operators must anticipate the composition of leachate that a site will generate and select the appropriate liner materials. The chemical resistance of any geomembrane materials, as well as LCR pipes, should be thoroughly assessed prior to installation.

Two of the most important considerations in the projects are the materials for construction, which are basically soil and geosynthetics and the geosynthetic component which is straighter to deal with because geosynthetics are man-made so some level of quality control can be exercised by the manufacturer.

The soil, however, is a naturally occurring material that is a bit more complicated than most people realize, and a basic understanding of the nature of clay is valuable in recognizing the potential effects that the material can have on the construction schedule (Fig. 2.32).

The soil group consisting of particles smaller than two (2) microns (0.002 mm or approximately 1/10,000 inch) is referred to as the term clay. The particular features of clay are related to the size and chemical properties of the particles that make up the soil. Hydraulic conductivity, which is necessarily the velocity of fluid passing through the soil and is normally stated as centimeters per second (cm/s) or feet per minute (ft/min), is a unique characteristic of clay soils that make them favorable for application to landfill construction. The clay is compacted to a specific standard in order to perform the stipulation as a measure assuring the possible favorable hydraulic conductivity in construction. The problem begins with reaching that standard. The amount of moisture in the soil is certainly related to the achievable density during compaction. Dry weight (pounds) of soil in a unit volume (cubic foot), or pounds per cubic foot (PCF) is the common expression for density. The ratio of the weight of water in a unit volume of soil to the weight of dry soil in that same unit volume, normally expressed as a percentage is the term used for moisture.

2.18.5   Construction Quality Assurance

A construction quality assurance (CQA) program should be followed for liner installation in order to diminish holes in a geomembrane liner (whether caused by product
defects, transportation, installation, or seaming) and to meet needed standards. Landfill owners carry out the CQA program which is a planned system of activities to
assure the construction of cells and related facilities according to the design. A facility
CQA program should be developed in the design stage and should be reviewed by
state regulatory agencies prior to issuing an allowance for construction.

2.18.6   Quality Assurance/Construction Documentation
Report

Authorization to use a new facility is typically unforeseen upon Departmental review
and approval of a quality assurance/construction documentation report.

An adequate report contains, at least, the following information (Vesilind et al.
2002a; Bagchi 2004; Christensen et al. 2005):

- As-built engineering drawings illustrating the following information:

  a. Completed sub-base elevations.
  b. Final liner grades.
  c. Top of drainage blanket grades.
  d. Leachate collection lines, clean-outs, and manholes with spot elevation every
      100 feet along the lines and at all manhole entrances and exits.
  e. Drainage characteristics.
  f. All monitoring devices.
  g. Spot elevations at all breaks and slope and on approximate 100-foot centers.
  h. All test locations.
  i. Other site information as appropriate.

- Engineering cross-sections, a minimum of one east–west and one north–south
  through the completed area.
- An extensive narrative describing how the construction of the project was
  performed the accompanying analysis of the soil, liner and any other testing data.
  An appendix containing all the raw data from the field and laboratory testing
  should be included in this report.
- A series of 35 mm color prints documenting all major aspects of the site
  construction.

A registered professional engineer should accredit the construction of the site to
ensure its completion according to the approved plans. Any deviations from the plan
should be specified and clarified.

2.18.7   Landfill Operations

The development of a workable operating plan, a filling plan for the location of solid wastes, an approximate of the equipment necessary, development of landfill operating records and billing information, a load inspection for hazardous waste, traffic control on highways leading to the landfill, and a site safety and security program are the important elements of a landfill operation plan. Another part of landfill management is a continuing community relations program. Effective disposal of the highest amount of waste in the engineered landfill along with the protection of the environment, workers, and customers are the main goals of landfill operations.

Landfill Operating Schedule

The followings are the factors required to be considered in developing operating schedules:

- Arrival sequences for collection vehicles
- Traffic patterns at the site
- The time sequence to be followed in the filling operations
- Effects of wind and other climatic conditions
- Commercial and public access.

Restriction of public access to the site until later in the morning may be required due to heavy truck traffic early in the morning, for instance. Moreover, to not stop the landfill operations in unusual weather conditions, a filling sequence should be devised due to different winter conditions. When winds exceed, for example, 35 mi/h, landfill closure may be required, if the control of blowing paper during high-wind conditions is not possible.

Landfill Operating Records

The requirement of an entering scale and gatehouse is due to the determination of the quantities of waste that are disposed of. The gatehouse will be used by personnel responsible for weighing the incoming and outgoing trucks. The number of vehicles that must be processed per hour and the size of the landfill operation is the factors affecting the complexity of the weighing facilities. Detection of the existence of radioactive substances in the incoming wastes is performed with radiation detectors which are added to weigh stations in larger landfills. Recording video systems are regularly observing many weighing stations. Figure 2.33 is represented some examples of weighing facilities. The monitoring of the performance of the operation and determination of the in-place specific weight of the wastes will be possible in the case of the defined weight of the solid wastes. Charging of participating agencies and private haulers for their contributions will be based on the weight records.

Filling Sequences

Daily deliveries of waste are located in lifts or layers on top of the liner and leachate collection system to depths of 20 m (65 ft) or greater at an active landfill. In areas

Fig. 2.33 Typical truck-weighing facilities at the small landfill

with a high groundwater table or subsurface conditions preventing excavation, the area or hill type of landfill construction is typically used (Pazoki et al. 2015b). Waste placement in cells or trenches dug into the subsurface is provided by the excavation or trench technique. The detached soil is usually applied as a daily, intermediate, or final cover.

Furthermore, waste can be located against lined canyon or ravine side slopes in suitable cases. Critical issues for this kind of waste placement are control of slope stability and leachate and gas release.

Waste in the first lift is chosen to avert heavy and sharp objects due to crucially protection of the leachate collection system during landfill operations. This layer is usually referred to as the operational layer. Compactor wheels are kept away from the leachate collection system through the placement of the waste in a special manner. Formation of the next layer takes place through filling in the next lifts moving outward generally in a corner. The filling sequence is designed at the moment of landfill design and permitting. The size of the working face should be sufficiently large, normally 12–20 ft (4–6 m) per vehicle in order to hold several vehicles unloading simultaneously. Compaction of the waste to the highest use of airspace is made possible using heavy equipment while the placement of the waste in the landfill. Refuse layer thickness, the number of passes made over the waste, slope (flatter slopes compact better by landfill compactors, steeper slopes (maximum 1:3) compact better by track-type tractors), and moisture content (wetter waste compacts more effectively than dry waste) are significant factors in the degree of expected compaction.

Working Area:

$A_{max} = (0.1\ W)/R$ (10.4).

A_{max}: the maximum working area (m^2).

R: the average annual rainfall (m).

W: the average annual waste input (metric tons).

The daily working area for waste housing should be kept as small as possible, said, no more than one hectare, at most sites.

Waste acceptance procedures and criteria

A landfill operator must be satisfied that the waste has his permit conditions, before acceptance of waste at a landfill site: the waste acceptance procedures (WAP) and waste acceptance criteria (WAC). The following procedures should be pursued in the case of the decision of disposal to landfill is the best arrangement option for your waste, or the waste would not be accepted.

– Waste acceptance procedures

The Council Decision speculates waste acceptance procedures (WAP). The site-specific WAP owned by landfill operators should be built from this foundation. Determination of waste suitability to go to the landfill, and if so, to which class is performed by The Council Decision WAP. Three steps for identification and regularly control the main features of the waste have consisted of The WAP:

Level 1: Basic characterization The waste composition and properties and therefore its suitability to be accepted and at which class, is required to be known, prior to sending a load to the landfill.

Level 2: Compliance testing A waste which is 'regularly arising', e.g. from an industrial process, it should be regularly audited to assure the steadiness of its properties.

Level 3: On-site verification Each delivery at the landfill should be controlled by the operator to verify its expectancy and be free of contaminations in storage or transport.

As well as WAP, all waste holders have a Duty of Care. This means that you must:

– take all feasible measures to prevent a breach of any legislation, including permit conditions;
– make sure that your waste is contained and does not go out of your control;
– pass on a written 'waste description' with the waste which others will need to avert a breach of legislation. The 'basic characterization' summarized above will implement this constraint, and
– pass on a waste 'transfer note' with the waste (includes the use of a season ticket where appropriate).

Application of the List of Wastes Regulations4 in the identification of wastes in transfer notes is considered necessary by The Environmental Protection (Duty of Care) Regulations 1991.

Waste Acceptance Criteria

Inert, non-hazardous and hazardous waste acceptance criteria are determined by the Council Decision. These are:

- a list of wastes which may be accepted at a landfill for inert waste without testing;
- limits on the leachability of specific parameters; and
- limits the organic content of the waste.

No numerical WAC limits on landfills are defined for non-hazardous waste.

In many examples, sampling and checking the wastes to see whether they are within the limits is necessary.

Banned wastes

The following types of waste cannot be sent to landfill sites:

- liquid waste;
- waste which in a landfill would be explosive, corrosive, oxidizing, flammable or highly flammable;
- hospital and other clinical wastes—from medical or veterinary organizations— which are infectious;
- chemical substances from research and development or teaching activities (such as laboratory residues) which are not identified or which are new, and whose effects on man and/or the environment are not identified;
- whole and tore down used tires—apart from tires used as an engineering material, bicycle tires, and tires having an outside diameter of more than 1,400 mm.

On-site verification

There are three levels of on-site verification:

- documentation check
- visual inspection
- periodic sampling.

Control of Birds

Birds at the landfill site are not merely a bother; strict challenges may be caused if the landfill is placed in the vicinity of an airport. Landfills should be constructed within 6 mi of an airport, and in some conditions in greater distances according to federal regulations. Application of noisemakers, recordings of the sounds made by birds of prey, and overhead wires are available techniques to control birds at landfill sites. Early 1930 is the time origin of controlling birds at reservoirs and fishponds with overhead wires. County Sanitation Districts of Los Angeles County in the early 1970s was the first user of overhead wires to control seagulls at landfills (see Fig. 2.34). Apparently, the wires interfere with the seagulls' guidance system because of the go down in a circular pattern when landing. The poles are commonly spaced 50–75 ft apart; with line spans from 500 to 1200 ft. Crisscrossing enhances the efficiency of the wiring system. Although stainless steel wire has also been employed, a typical 100-lb test monofilament fish line is used.

Fig. 2.34 Aesthetic considerations in landfill design: **a** view of the landscaped landfill in which filling operations are not visible from the nearby freeway; **b** overhead wire system used to control seagulls at landfills; **c** wire screen used to control blowing; **d** bulldozer to flatten the top layer

Control of Blowing Materials

Windblown paper, plastics, and other waste may cause a problem at some landfills with dependence on the location. The application of handy screens near the operating part of the landfill is a commonplace solution. With daily removal of the accumulated material on the screen, the problems with vectors are avoided. Maintenance of a good image of landfill operation is reached through efficient pickup of paper not being retained by portable screens. Tarpaulin should be used to cover the open-topped vehicles of waste harvesting to the landfill, in order to prohibit falling of paper and dust onto the highway. Either the vehicle operator or the landfill personnel should instantly pick up the waste falling out of a truck. To improve community relations, it is an efficient approach for landfill operators to periodically collect all litter along the roads heading to the landfill.

Control of Pests and Vectors

Pests, like mosquitoes and flies and rodents, such as rats and other burrowing animals are the major considerable vectors in the arrangement and operation of landfills. Placement of daily cover and elimination of steady water controls flies and mosquitoes. In areas where white goods and used tires are stored for recycling, stagnant water may be a problem. Most of the problems are avoided using covered facilities for the stockpiling of these materials. The application of daily cover eliminates the problem of rats and other burrowing animals (see Fig. 2.34d).

Public Health and Safety
Public health and safety matters are connected to the health and safety of the public and also workers.

Health and Safety of Workers
A critical factor in the operation of a landfill is the health and safety of the workers. Puncture wounds from sharp objects, equipment rollovers, laborers being run over, personnel falling into holes, and asphyxiation in confined spaces are among the types of occurring accidents at landfills. Requirements for an extensive health and safety program for the workers at landfill sites have been devised by the federal government, through the Occupational Safety and Health Administration (OSHA) and state-instituted OSHA-type programs. The newest regulations should be followed in the development of worker health and safety programs due to continual changes in these programs. Types of protective clothing and boots, air-filtering headgear, and puncture-proof gloves supplied to the workers should be carefully noticed depending on the activities at the landfill.

Safety of the Public
Landfill operators have been obliged to recheck past operations regarding public safety and site security due to safety concerns and the many new limitations governing the operation of landfills, as previously mentioned. Diminishing public contact with the working operations of the landfill through the application of an efficient transfer station at the landfill site is getting popular, consequently.

Site Safety and Security
Landfill operators have been forced to enhance security at landfill sites significantly due to the rising number of lawsuits over accidents at landfill sites. "No trespassing" and other warning signs have been posted and access to the landfill has been restricted through fencing in most sites, recently. In some locations, monitoring of landfill operations and access has been performed using television cameras.

Community Relations
Preservation of good community relations is an important and usually neglected aspect of landfill operation. A dialogue should be maintained between the land-fill manager and neighbors, municipal leaders, community activists, and state government representatives to build trust through honest communications. Although continual support for the landfilling operation is not secured through community relations activities, however, poor communications will definitely result in complaints and problems.

Landfill Operator Training
The operation of landfills must be accompanied by a careful plan of the employee training program as it is considered being recognized. If the landfill which contains many technical components of a unique nature is not suitably managed, it will not obtain its favorable results. The operation of a landfill is a costly activity. Even higher

expenses and investment loss will result due to system failures as a consequent of improper management. Environmental objectives, the design elements of the landfill, the equipment operation, environmental regulations, and health and safety protection, are the target of the training programs for operators. Possession of a valid license for the landfill manager and possibly the equipment operators achieved after certain training programs is a requirement by many governmental units. In some jurisdictions, lasting education is also necessary on a continuing basis.

Landfill Operations Manager

1. In charge of all matters concerned with the development, operation, and completion of the landfill in accordance with the disposal plan. Primary duties include the following:
2. Forward planning for human, technical, and financial resources for the appropriate operation and maintenance of the landfill.
3. Preparation of such forecasts for consideration by the municipality's officer responsible for waste management.
4. Enrollment of staff and procurement of equipment and materials within approved budgets to operate and maintain the landfill.
5. Development of a detailed health and safety plan, emergency plan, and environmental monitoring plan as extensions to the disposal plan.
6. Assignment of tasks to site staff and assuring that the landfill supervisor(s) authority is known to others. Training staff as necessary.
7. Instruction of landfill supervisor(s) on a frequent basis as to the areas of the site to be prepared/filled/capped, and the degree of environmental monitoring to be implemented.
8. Review of daily and weekly reports prepared by the landfill supervisor(s), and others, on-site activities.
9. Preparation of monthly/quarterly/annual management information reports, approval of salary payments, and authorization of equipment and materials purchases.

Landfill Supervisor

Responsible for the daily activities on the landfill site according to instructions received from the landfill operations manager. Duties include (Vesilind et al. 2002a; Bagchi 2004):

1. Supervision of labor to control the admission of wastes, movements of vehicles within the site, tipping in approved areas, compaction, and covering of tipped waste.
2. Maintenance and supervision of plant and/or vehicles.
3. Winning borrow material for use as daily/intermediate/final cover and stockpiling of cover material.
4. Supervision of housekeeping activities on site carried out by nontechnical staff.
5. Preparation of daily and weekly activity reports.

6. Reporting to the landfill operations manager any developing problems in the areas of personnel, equipment, materials, waste inputs, public complaints.

2.19 Monitoring

Landfill monitoring is demanding for the operation of a landfill. Leachate head on the liner, leakage through the landfill liner, groundwater quality, ambient air quality (to ensure compliance with the Clean Air Act), gas in the surrounding soil, leachate quality and quantity, landfill-gas quality and quantity, and stability of the final cover are commonly monitored by landfill operators. The capabilities to maintain a properly designed leachate-collection system, monitor head on the liner, store or dispose of leachate outside of the landfill, and eliminate leachate at proper rates are required in controlling of a head on the liner. A lysimeter, such as that depicted in Fig. 2.35 is often used to detect leakage through a single liner (Tchobanoglous et al. 1993). The landfill design is the basis of the locations and number of lysimeters; however, assurance of repetition is achieved through provision of more than one. The maximum head founded at the point of maximum leakage potential below the crest of the landfill is the place of lysimeter.

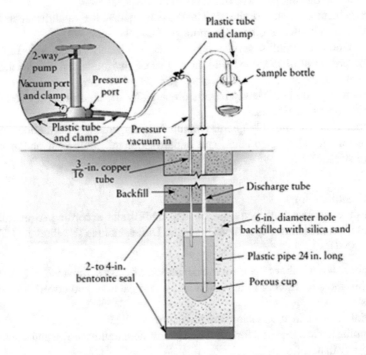

Fig. 2.35 Porous cup suction lysimeter for the collection of liquid samples from the landfill

The building and sampling of monitoring wells near the landfill is the general approach for groundwater monitoring. The evaluation of groundwater quality at multiple depths is provided by the placement of a series of wells in each location, which is an optimum arrangement. The impact of the landfill on groundwater is determined by locating the good clusters up-gradient and down-gradient from the landfill to evaluate background groundwater quality. Further monitoring wells are also located at the property borders. As designated in RCRA Subtitle D regulations, various organic and inorganic groundwater constituents are determined through the quarterly sampling of the wells. More thorough monitoring is required, when down-gradient constituents have a statistically significant increase in concentration as compared with up-gradient concentrations. More expanded monitoring and remedial action should be planned if down-gradient levels ongoing exceed drinking water standards.

In the case of demonstrated no potential for the migration of hazardous constituent from the landfill to the uppermost aquifer during the active life of the landfill and the post-closure period, the cease of groundwater monitoring will be permitted by the EPA. Generally, the application of computer models for simulation of contaminant movement in the subsurface helps to perform this demonstration. EPA especially devised The Multimedia Exposure Assessment Model (MULTIMED) for this purpose. Not rising the methane gas concentration to 25% of the lower explosive limit for methane (5% concentration) in facilities and to the lower explosive limit at the landfill border should be assured by the landfill operators. Landfill owners will be able to test landfill gas by the placement of gas monitoring probes, such as the one shown in Fig. 2.36, at the property borderlines and different other locations around the landfill site. Collection of any landfill gas migrating towards the property borders takes place by locating the gas extraction wells, as shown in below, at the property boundaries and other locations. The location of the probes is determined

Fig. 2.36 Gas extraction well in place

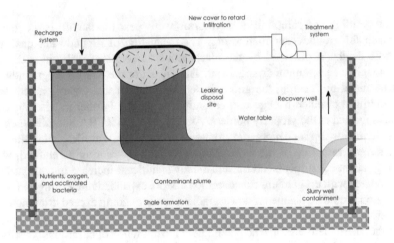

Fig. 2.37 Typical examples of a groundwater remediation system involving the use of slurry wall containment, recovery wells, treatment of contaminated groundwater, and groundwater recharge with nutrient addition to achieving in situ remediation

with concern to the hydrogeologic properties of the site, soil conditions, and the placement of landfill structures.

2.20 Landfill Closure, Post-closure Care, and Remediation

The terms used to explain the operations done to completed landfill in the future are landfill closure and post-closure care. The operator of a landfill should save enough money to close, maintain and monitor the facility properly for 30–40 years after its completion according to states and the federal government legislations to ensure the maintenance of the completed landfills in the 30–50 years in the future. In some European countries, greater future care periods are considered. The access to a closure plan clearly outlining the closure requirements is the most important factor in the long-term maintenance of a completed landfill. A design for the landfill cover and the landscaping of the completed site should be included in the closure plan. Furthermore, long-term plans for runoff control, erosion control, gas and leachate collection and treatment, and environmental monitoring should be considered in the closure.

Cover and Landscape Design

Redirecting surface runoff and snowmelt from the landfill site and supporting of the landscaping design selected for the landfill are the requirements of a landfill cover. Progressively, local plant and grass species, as opposed to imported plant and grass species, are the basis for the ultimate landscaping arrangement. Landscaping of the desert type is preferred in many locations in the southwest with water shortage.

Control of Landfill Gases

A major concern in the future maintenance of landfills is the control of landfill gases. There is a type of gas control system installed prior to the landfill completion in most modern landfills due to concern about the uncontrolled release of landfill gases. Retrofitting with gas collection systems is the operation performed in older completed landfills deprived of gas collection systems.

Collection and Treatment of Leachate

Another important concern in the long-term maintenance of landfills is the control of leachate discharges. As previously noted, there is a leachate control system in most modern landfills. Older completed landfills with no leachate collection systems are being retrofitted with leachate collection systems. There is a similarity in these retrofitted collection systems in construction with vertical gas wells in which leachate pumps are installed. Installation of the leachate head wells is difficult because of blockage faced during drilling. Their efficiency may be confined due to poor hydraulic flow conditions through the waste.

Post-closure Care

The normal checkup of the completed landfill site, maintenance of the infrastructure, and environmental monitoring are included in the post-closure care plan. These subjects are considered briefly as follows.

Routine Inspections

In order to progressively monitor the condition of the completed landfill a normal inspection program must be founded. When remedial action should be performed criteria must be devised to determine the type of action. For example, how much settlement will be permitted prior to regarding must be undertaken?

Environmental Monitoring Systems

Assurance of no release of pollutants from the landfill possibly affecting health or the surrounding environment is achieved through long-term environmental monitoring. (1) Vadose zone monitoring for gases and liquids, (2) groundwater monitoring, and (3) air quality monitoring are involved in the required monitoring at completed landfills. The regulations of the local air pollution and water pollution control agencies determine the number of samples collected for analysis and the frequency of collection. A baseline procedure for a sampling of groundwater that should be reviewed has been devised by the EPA (Vesilind et al. 2002a; Bagchi 1994, 2004).

Remediation

In the case of detection of higher levels of environmental release in the post-closure monitoring program, corrective actions may be required. Landfill gas migration, toxic air emissions, leachate polluting the groundwater, or some other unforeseen event are the cause of remedial actions. The strength of the corrective action and the long-term cost will be determined by the importance of the problem (Bagchi 2004; Pazoki et al. 2018).

Migration Control

Methane concentrations cannot be higher than 5% methane at the property boundary of the MSW landfill according to federal regulations. Even lower concentrations are required in some states. Extension of landfill gas migration to areas containing settlements may unexpectedly occur. The first steps necessary to start with no delay are securing the area and evacuation of the buildings through emergency measures. The presence of methane in buildings can be determined using proper equipment usually available at local fire departments. The purposes of installation of wells on or near the landfill and also settlements are stopping gas migration away from the site and elimination of the gas from the ground. The wells in or near to the landfill are operated for years until the concentration of the generated methane, not a danger. The wells located near the occupied buildings will occasionally operate, often on the order of months, until the methane in the vadose zone is reduced to safe levels (Tchobanoglous and Theissen 2005; Bagchi 1994).

Toxic Air Emissions

Installation of landfill gas control and recovery systems is surprisingly found critical by a number of landfill operators in order to confine the release of toxic compounds into the atmosphere. The technology and system configurations are explained previously in this chapter. The required period for operating these systems is unknown.

Groundwater Remediation

A harmful impact on groundwater quality is most possible from unlined landfills and landfills without leachate collection systems. If the pollution in groundwater monitoring wells is detected, a corrective action program is planned regarding federal or state regulations. The reduction of water drainage through the waste as a result of the placement of a new, highly impermeable cap over the landfill is the primary corrective step. Limitation or blockage of the movement of contaminated groundwater away from the landfill by the installation of bentonite slurry walls or the operation of recovery wells controlling subsurface hydraulics is the purpose of following designed measures. Treatment of the polluted groundwater in the aquifer in the vicinity of the site above-ground facilities and either re-injection, spraying onto nearby land, or discharge to surface water are the available processes. Elimination of contaminants from the groundwater may also be performed through in situ bioremediation techniques. In order to decrease future costs, remediation by natural dilution, where adventive, dispersive, adsorptive and biodegradation processes are involved are preferred to disinfect groundwater contamination. The time needed to reach an adequate level of groundwater quality enhancement is estimated using computer models simulating natural remediation. The completion of recovery measures by any of the remediation processes will take years or decades.

General Principles

Restoration plans are required to take into account the following matters (Tchobanoglous and Theissen 2005; Tchobanoglous et al. 1993; Bagchi 1994):

- The type of final cover (cap) for the completed landfill
- The interception of leachate from the site to avoid contaminating surface and groundwater
- The types of surface and groundwater monitoring which can be reached
- The new works and maintenance are necessary to continue to keep surface water away from the deposited waste.
- The methods to prevent soil erosion from the final cover.
- The options in access to maintain, or install, landfill gas and leachate collection (and treatment) systems
- The requirements necessary to maintain the long-term integrity of the final cover, to control establishments and provide re-vegetation
- The means to confine access to the site after closure and capping and the site's potential after-uses.

2.21 After-uses

In the vicinity of metropolitan areas, there may be strong competing pressures to return landfills to an obviously normal land profile. Potential after-uses include the following (Tchobanoglous and Theissen 2005; Bagchi 1994):

- Agriculture arable land, grazing, exercise pasture
- Forestation woodland, tree screens, nature reserves
- Amenity open space, buffer zones, airport runways
- Recreation parks, playing fields, sports complexes, tracks, and golf courses
- Habitation caravan sites, gardens, play areas, squatter (pre-urban) settlements
- Industry open-storage areas, parking, fabrication areas
- Among the important limitations affecting the application of a former landfill are those that arise from
- Low load-bearing capacity
- Extensive settling (especially uneven settling)
- The presence of combustible and potentially explosive gases
- The corrosive character of the decomposition products to concrete and steel, and the different biochemical internal landfill environment in general.

These limitations continue long after the fill has been finished. Climate (rainfall, temperature), the nature of the buried wastes, and the design and operational features of the landfill are the affecting factors in the duration of this aftercare period. For instance, in a country in a humid, tropical setting it may last 10, 20, or 30 years and more than 100 years in a desert environment. Three general classes, open space/recreation, agricultural, and urban development, can describe the final usage of completed landfills.

2.22   Landfill Mining

Digging up old landfills, separation of the non-biodegradable fraction, and applying the dirt and organic soil as a cover material for present landfills may be possible in the case of significant biodegradation occurred in the old landfill. Seemingly it is suitable for communities needing new landfills and landfill mining appeared to be highly practiced. For example, a south Florida landfill mined an old landfill and recovered some metals (Basak 2004) However, as the regulations of the landfill are getting gradually severe and difficult control of landfill gas, mining landfills is very cumbersome. The rainwater is allowed to make polluted runoff due to escaped gas to the atmosphere permitted through opening the cap of a landfill. Ultimately, due to dirt and difficult cleansing of mined materials, there is marginal cost. If landfill mining creates a new volume for extending the life of a present landfill and if the operation is done with a minimum environmental cost, it will be economically feasible. Since the only landfill tending to be fully biodegraded is the one which is flat and wet with few vertical lifts, it is a candidate for landfill mining. Excavation of the buried refuses and analyzing it to separate out the useful new cover materials are included in the landfill mining procedure. The non-biodegradable parts are of little value and are re-landfilled, normally. The time the landfill is closed will be significantly extended if the operation is successfully performed and therefore economical alternatives will be produced to set up new landfills.

References

Allen A (2001) Containment landfills: the myth of sustainability. Eng Geol 60(1–4):3–19

Bagchi A (1994) Design, construction, and monitoring of landfills

Barlaz MA, Milke MW, Ham RK (1987) Gas production parameters in sanitary landfill simulators. Waste Manage Res 5(1):27–39

Basak S (2004) Landfill site selection by using geographic information systems. www.metu.edu.tr

Bolton N (1995) The handbook of landfill operations: a practical guide for landfill engineers, owners, and operators. Blue Ridge Services

Chambers JC, McCullough MS (1995) From the cradle to the grave: an historical perspective of RCRA. Nat Res Environ 10(2):21–74

Christensen TH, Cossu R, Stegmann R (eds) (2005) Landfilling of waste: leachate. CRC Press

Delarestaghi RM, Ghasemzadeh R, Mirani M, Yaghoubzadeh P (2018) The comparison between different waste management methods of Tabas city with life cycle assessment assessment. J Environ Sci Stud 3(3):782–793

Bagchi A (2004) Design of landfills and integrated solid waste management. Wiley

Edgers L, Noble JJ, Williams E (1992) A biologic model for long term settlement in landfills. In: Mediterranean conference on environmental geotechnology, pp 177–184

Edil TB, Ranguette VJ, Wuellner WW (1990) Settlement of municipal refuse. In: Geotechnics of waste fills—theory and practice. ASTM International

El-Fadel M, Shazbak S, Saliby E, Leckie J (1999) Comparative assessment of settlement models for municipal solid waste landfill applications. Waste Manage Res 17(5):347–368

Ghasemzadeh R, Pazoki M, Hoveidi H, Heydari R (2017) Effect of temperature on hydrothermal gasification of paper mill waste, case study: the paper mill in North of Iran. J Environ Stud 43(1):59–71

Ghasemzade R, Pazoki M (2017) Estimation and modeling of gas emissions in municipal landfill (Case study: Landfill of Jiroft City). Pollution 3(4):689–700

Greenberg M, Hughes J (1992) The impact of hazardous waste Superfund sites on the value of houses sold in New Jersey. Ann Reg Sci 26(2):147–153

Kumar S, Chiemchaisri C, Mudhoo A (2011) Bioreactor landfill technology in municipal solid waste treatment: an overview. Crit Rev Biotechnol 31(1):77–97

Pazoki M, Delarestaghi RM, Rezvanian MR, Ghasemzade R, Dalaei P (2015a) Gas production potential in the landfill of Tehran by landfill methane outreach program. Jundishapur J Health Sci 7(4)

Pazoki M, Pari MA, Dalaei P, Ghasemzadeh R (2015b) Environmental impact assessment of a water transfer project. Jundishapur J Health Sci 7(3)

Pazoki M, Abdoli MA, Ghasemzade R, Dalaei P, Ahmadi Pari M (2016) Comparative evaluation of poly urethane and poly vinyl chloride in lining concrete sewer pipes for preventing biological corrosion. Int J Environ Res 10(2):305–312

Pazoki M, Ghasemzade R, Ziaee P (2017) Simulation of municipal landfill leachate movement in soil by HYDRUS-1D model. Adv Environ Technol 3(3):177–184

Pazoki M, Ghasemzadeh R, Yavari M, Abdoli M (2018) Analysis of photocatalyst degradation of erythromycin with titanium dioxide nanoparticle modified by silver. Nashrieh Shimi va Mohandesi Shimi Iran 37(1):63–72

Pfeffer JT (1992) Solid waste management engineering. Prentice Hall

Pohland FG (1996) Landfill bioreactors: fundamentals and practice. Water Qual Int 9(1996):18–22

Renou S, Givaudan JG, Poulain S, Dirassouyan F, Moulin P (2008) Landfill leachate treatment: review and opportunity. J Hazard Mater 150(3):468–493

Rushbrook P, Pugh M (1999) Solid waste landfills in middle and lower-income countries: a technical guide to planning, design, and operation. The World Bank

Sahadewa A, Zekkos D, Lobbestael A, Woods RD (2011) Shear wave velocity measurements at municipal solid waste landfills in Michigan. In: Proceedings of 14th Pan-American conference on soil mechanics and geotechnical engineering. Toronto, Canada

Schwartz EM (2001) A simple approach to solid waste planning for urbanizing counties. Doctoral dissertation, Department of Geosciences. University of Missouri--Kansas City

Şener B, Süzen ML, Doyuran V (2006) Landfill site selection by using geographic information systems. Environ Geol 49(3):376–388

Shayesteh AA, Koohshekan O, Khadivpour F, Kian M, Ghasemzadeh R, Pazoki M (2020) Industrial waste management using the rapid impact assessment matrix method for an industrial park. Global J Environ Sci Manage 6(2):261–274

Tang WH, Gilbert RB, Angulo M, Williams RS (1994) Probabilistic observation method for settlement-based design of a landfill cover. In: Proceedings of the conference on vertical and horizontal deformations of foundations and embankments. Part 2 (of 2). Published by ASCE, pp 1573–1589

Tchobanoglous G, Theisen H, Eliassen R (1993) Engineering principles and management issues. Mac Graw-Hill, New York, p 978

Tchobanoglous G, Theissen H, Vigil S (2005) Integrated Solid Waste Management

Turner JH (1996) Off to a Good Start: The RCRA Subtitle D Program for Municipal Solid Waste Landfills. Temp Envtl L & Tech J 15:1

Vesilind PA, Worrell WA, Reinhart DR, Vesilind A (2002a) Solid waste engineering. Brooks, Cole Pacific Grove

Vesilind PA, Worrell W, Reinhart D (2002b) Municipal solid waste characteristics and quantities in solid waste engineering. Thomson Learning Inc., Singapore

Wagner TP (1999) The complete guide to the hazardous waste regulations: RCRA, TSCA, HMTA, OSHA, and Superfund. Wiley

Watts KS, Charles JA (1999) Settlement characteristics of landfill wastes. Proc Inst Civil Eng-
    Geotech Eng 137(4):225–233
Zekkos D, Flanagan M (2011) Case histories-based evaluation of the deep dynamic compaction
    technique on municipal solid waste sites. In: Geo-frontiers 2011: advances in geotechnical
    engineering, pp 529–538

Chapter 3
Leachate Quality

3.1 Introduction

Leachate refers to every liquid material which takes out solutes, suspended solids or any ingredient of the material along which it has passed from solid waste. Literally, leachate is a common term in the bio-sciences where it conveys the special meaning of a liquid material that has solved or drawn biologically detrimental materials which can probably return to the environment, again. It is widely used in landfilling of material that is liable to decay or industrial waste (Tchobanoglous et al. 1993).

However, to put it in more detail, leachate refers to every liquid material that draws from the ground or cumulative material and encompasses a significantly high level of biologically adverse material extracted from the material that it has infiltrated (Pazoki et al. 2015a).

Leachate can be categorized into two sub-categories, primary leachate, and secondary leachate (Christensen et al. 2005).

3.1.1 Primary Leachate

Leachate may be the product of more than land capacity moisture of waste. That is, primary leachate is the initial result of the first process after burying the waste; indeed, it is more than necessary moisture of waste in the land.

Commonly, the production process begins with pouring waste into the cells. The very first waste moisture is the initial source of raw leachate which emanates from the waste and takes the form of waste.

After being formed above the primary compact (and subsequent compression of liable to decay waste) in a landfill, the land capacity will continuously decrease. Such land capacity decrease is important in terms of leachate production since the fast pace of decrease in the land capacity for the primary compression where the

© Springer Nature Switzerland AG 2020
M. Pazoki and R. Ghasemzadeh, *Municipal Landfill Leachate Management,*
Environmental Science and Engineering,
https://doi.org/10.1007/978-3-030-50212-6_3

waste has proportionally great humidity (e.g. Iran) gives rise to more than necessary waste moisture after compression to its outlet from. The above mentioned was the detailed production procedure of primary leachate.

3.1.2 Secondary Leachate

A downfall in almost every weather conditions results in water penetration into the landfills of waste. Moisture penetrating the landfill beneath its own gravity goes down and is soaked up where waste and materials reach together. Wastes that are in solid form are given a dehydrating capacity in terms of materials and components and their compression level in the landfill and also given the land capacity.

Water components moved by buried waste is commonly less than soaking level.

Such a breach in the capacity will lead to sucking up water penetrated into the waste. At the same time when buried materials touch the field capacity, (after being soaked) the residual moisture which has not been taken up by solid waste turns into leachate.

The leachate that has been resulted from this procedure is known as secondary leachate.

3.2 Leachate Production Process

3.2.1 Aerobic and Anaerobic Biochemical Processes

Aerobic decay process: The initial step of aerobic decay of organic compounds is usually a short process because wastes require a high level of oxygen since there is a confined amount of oxygen available inside landfills (Canziani and Cossu 2012).

The single landfill layer associated with aerobic transformation is the upper landfill level since oxygen is caught with fresh waste and is provided by downfall and rainfall. At this step, it can be seen that protein contents are transformed into amino acids, then into carbon dioxide, water, nitrates, and sulfates. These are common catabolism products of all oxygen-borne processes.

Carbohydrates are transformed into CO_2. Water and fats undergo hydrolysis and turn into fatty acids and glycerol. After that, they are transformed into simple catabolism products by the moderate process of volatile acids and alkalis formation. Cellulose, which is composed of many organic portions of wastes, is decayed by being subjected to outside-cell enzymes into glucose which is then taken up by bacteria and changed to CO_2 and water. This phase, because of heat-releasing processes of biological oxidation, may go up in temperatures if the waste is not compressed. Typically, the oxygen-borne process is relatively limited in duration and results in no significant leachate production (Christensen et al. 2005; Canziani and Cossu 2012).

In aged landfills, when only the more difficult organic carbon is left in the landfill of wastes, a second oxygen borne process may take place in the near-surface level of landfill. During this process, the generation of methane is so low that air will start spreading out of the atmosphere, resulting in aerobic zones and zones with oxidation–reduction potentials that are too much for the generation of methane (Christensen and Kjeldsen 1989).

Anaerobic degradation processes: 3 distinct processes can be recognized in the anaerobic waste decay. Acid fermentation is known as the initial anaerobic decay process which reduces the pH content of leachate, increases volatile acidconcentration and results in considerable concentrations of inorganic ions (e.g.: Cl^-, SO_4^{2-}, Ca_2^+, Mg_2^+, Na^+) (Christensen and Kjeldsen 1989; Pazoki et al. 2016).

When oxidation–reduction potential falls downs, metal sulfideswith low quality of being solved are slowly produced, precipitate iron, manganese,, and other heavy metals that were solved by being subjected to acid-fermentation, the first high content of sulfates may de decreased with a low pace.

High generation of volatile fatty acids and high partial pressure of CO_2 results in reduced pH content. The increased concentration of negatively and positively charged ions is again the product of percolating easily soluble material composed of original waste materials and decayed products of organic components. The first anaerobic processes are provoked by a series of blended anaerobic microbes, made up of purely anaerobic bacteria and facultative anaerobic bacteria (Christensen and Kjeldsen 1989; Ghasemzade and Pazoki 2017).

The facultative anaerobic bacteria slow down oxidation–reduction capacity and thus methanogenic bacteria findan opportunity to grow. Indeed, methanogenic bacteria are responsive to the existence of oxygen and need an oxidation–reduction potential below −330 mV. The characteristic of leachate generated during this process is high BOD5 values (typically >10.000 mg/l), high BOD5/COD proportion (typically > 0, 7), acidic pH values (usually 5–6), and ammonia (mostly 500–1.000 mg/l); especiallyy, the latter because of hydrolysis and fermentation of components with high content of protein. The second intermediary anaerobic phase (Phase III, Fig. 3.1) begins with a low formation pace of methanogenic bacteria. This formation process may be hindered by the surplus of organic volatile acids since they are toxic to methanogenic bacteria when available at concentrations of 6.000–16.000 mg/l.

The dosage of methane in the gas rises, but hydrogen, CO_2 and volatile fatty acids fall down. Added to this, a dosage of sulfate decreases as a result of the biological reduction. Fatty acid transformation leads to the increased level of pH and therefore alkalinity with lower insolubility of calcium, iron, manganese and heavy metals. The latter is possibly deposited in solid form as sulfides. Ammonia particles are freed but not transformed when subjected to the oxygen-less environment (Canziani and Cossu 2012; Christensen and Kjeldsen 1989).

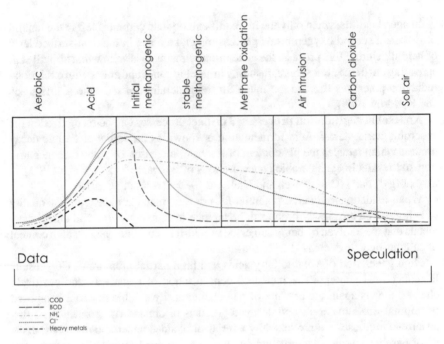

Data

Speculation

COD

BOD

NH₄⁺

Cl⁻

Heavy metals

Fig. 3.1 Illustration of developments in leachate and gas in a landfill cell (Christensen and Kjeldsen 1989). Phase I: Aerobic phase, Phase II: Acid phase, Phase III: Intermediate methanogenic phase, Phase IV: Stabilized methanogenic phase, Phase V: Final aerobic phase

The 3rd characteristic of anaerobic decay is methanogenic fermentation provoked by methanogenic bacteria. The pH range sustained by methanogenic bacteria is highly confined and extends between 6 and 8. At this phase, the structure of leachate is signified by mostly neutral levels of pH, low dosage of volatile acids and wholly solved solids while biological gas produces typically higher than 50% methane content. This proves that most of the organic material solubility has decreased during the landfilling process; however, the process of stabilizing waste will go on for so many years (Canziani and Cossu 2012; Worrell and Vesilind 2011).

The characteristic of leachates generated in this process is proportionally low biochemical oxygen demand level and low proportion of biochemical oxygen demand to chemical oxygen demand. All over the first acetogenic process, ammonia particles are released. In addition, the extent of leachate dosage, with regard to decaying of some relevant parameters, is shown in Table 3.1. What is more, in 1990, collected leachate dosages from German landfills during the 70th and 80th. With respect to this measurement, the organic materials (COD, BOD5, TOC) and AOX, SO₄, Ca, Mg, Fe, Mn, Zn, and Cr are specified by the biological and chemical processes of

Table 3.1 Chemical composition of landfill Leachate

Parameter	Median
COD (mg/l)	150–100,000
BOD5 (mg/l)	100–90,000
pH	5.3–8.5
Alkalinity (mgCaCO$_3$/l)	300–11,500
Hardness (mgCaCO$_3$/l)	500–8900
NH$_4$ (mg/l)	1–1500
Norg (mg/l)	1–2000
Ntot (mg/l)	50–5000
NO$_3$ (mg/l)	0.1–50
NO$_2$ (mg/l)	0–25
Ptot (mg/l)	0.1–30
PO$_4$ (mg/l)	0.3–25
Ca (mg/l)	10–2500
Mg (mg/l)	50–1150
Na (mg/l)	50–4000
K (mg/l)	10–2500
SO$_4$ (mg/l)	10–1200
Cl (mg/l)	30–4000
Fe (mg/l)	0.4–2200
Zn (mg/l)	0.05–170
Mn (mg/l)	0.4–50
CN (mg/l)	0.04–90
AOX[a] (μgCl/l)	320–3500
Phenol (mg/l)	0.04–44
As (μg/l)	5–1600
Cd (μg/l)	0.5–140
Co (μg/l)	4–950
Ni (μg/l)	20–2050
Pb (μg/l)	8–1020
Cr (μg/l)	30–1600
Cu (μg/l)	4–1400
Hg (μg/l)	0.2–50

[a] Adsorbable organic halogen

the landfill and there are notable differences between their levels in acid phase and methanogenic process. Added to this, studying 33 landfills in Northern Germany showed that leachate dosage is mostly taken from the late 80th and early 90th. Based upon the proportion of biochemical oxygen demand to chemical oxygen demand, the research showed 3 main periods: BOD5/COD = 0.4 Intermediate phase: 0.4 > BOD5/COD > 0.2 Methanogenic phase: BOD5/COD = 0.2. Organic parameters of the two investigations are significantly different (Christensen and Kjeldsen 1989; Pazoki et al. 2017).

Furthermore, findings of research on younger landfills leachate indicate that dosages of chemical oxygen demand, biochemical oxygen demand 5 and total organic carbon are less than those reported 10 years ago. These findings may be justified through developed technologies in the waste landfilling process. In many of the recent landfills, waste buildup and accumulation in less dense layers along with an oxygen-less pretreated lower layer was accomplished. This resulted in a shortened period of acid phase and an accelerated transformation of organic leachate materials to the gassy phase, the decay and transformation of organics to methane and carbon dioxide (Qasim and Chiang 1994).

3.3   Chemical Features of Leachate

The leachate content presented here is derived from more than 70 urban solid waste hygienic landfills in Europe and the U.S.A. Added to this, sampling, maintenance and pretreatment processes as well as analytical methods used to describe leachate may also have significant impacts (Christensen and Kjeldsen 1989).

Table 3.1 shows dosage limits for the major parameters of leachate in the landfill (Ehrig 1989; Andreottola et al. 1990). Figures 3.2, 3.3, 3.4, and 3.5 show the orientation of more outstanding parameters with regard to landfill age (Andreottola et al. 1990). Time orientation of organic material dosages (BOD5, COD) audibly shows that transition from the acid phase to methanogenic phase results in a significant reduction in dosage. Likewise, pH indicates an increasing trend based upon the evolution of biochemical reaction explain above.

The orientation of organic materials are presented in Fig. 3.6, leachate has been drawn against Landfill, represented as a rate of Total Organic Carbon (TOC) and this figure indicates that the generation of volatile acids, correlating with the first phase of anaerobic decay, is prevailing in the early period of landfill formation (Andreottola et al. 1990; Chian and DeWalle 1977).

The hydroxyl aromatic combination that is available in humic and fulvic-borne breaks of leachate organics demonstrate a partial reduction pursuant to landfill age yet indicate the greater ratio of TOC detected in the aged landfill (Andreottola et al. 1990; Chian and DeWalle 1977).

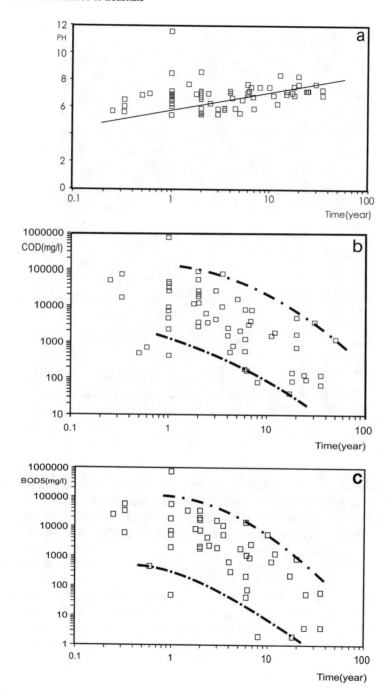

Fig. 3.2 Trends of pH (**a**), COD (**b**) and BOD5 (**c**) in comparison with landfill age for different landfill leachates

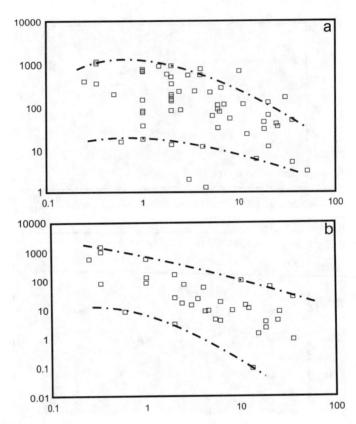

Fig. 3.3 Trends of NH$_4$ (**a**) and Norg (**b**) in comparison with landfill age for different landfill leachates

Leachate that follows the hydrolysis and fermentation of the protein breaks of the biologically decomposable underlying layer has high ammonia content. Ammonia content is likely to reduce gradually in accordance with the hypothetical orientation explained in Fig. 3.1 at the beginning of the second intermediate anaerobic stage. The orientation for organic nitrogen and the likeliness to reduce can be seen in Fig. 3.3. Added to this, Fig. 3.4 indicates the dosages of sulfates and chlorides in leachate as a function of time (Andreottola et al. 1990).

Also, concentrations that can be seen in a recently formed landfill are high with respect to sulfates. Sulfides reduction in an oxygen-less environment enforced quantized sulfate decrease in the landfill. Sulfide production may enforce sedimentation by different heavy metals available in the leachate. A great amount of primary dosage of chlorides reduces as the landfill becomes older due to the washing event. Dosages

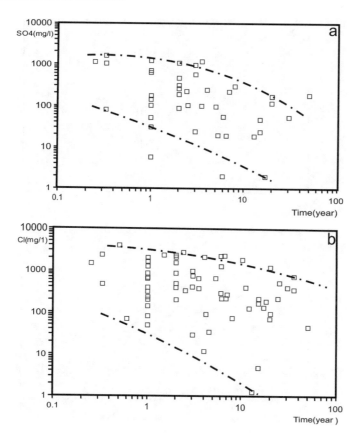

Fig. 3.4 Trends of SO$_4$ (**a**) and Cl (**b**) in comparison with landfill age for different landfill leachates

of iron, zinc, and manganese, selected as the more indicative of metals available in leachate are reported in Fig. 3.5. During the first stage of landfill formation, there is a high level of metal solubilization due to low levels of pH resulted from the high generation of pH gives rise to a lower level of solubilization as depicted in the Fig. 3.5 orientation report.

Trace metals' lixiviation is affected by the stabilization of organic compounds. Many organic materials with nitrogen, oxygen and sulfur content can form dissolved coordination compounds and thus increase the metal dosages.

Humic and fulvic acids are regarded as intense complexing ligands and may play a significant role in the long-time release of heavy metals from landfills. Also non-organic ligands, especially, chloride, may result in a coordination compound.

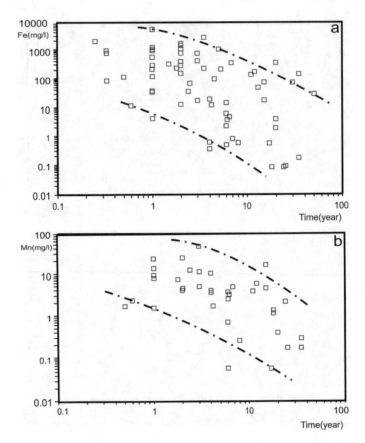

Fig. 3.5 Trends of Fe (**a**) and MN (**b**) in comparison with landfill age for different landfill leachates

The phenomena of adsorption and coordination are possibly highly liable for the debilitation and organization of trace metals in the accumulation of waste. Sorption may control the dosage of a structure to below the dosage set by chemical sedimentation subjected to oxidation properties. Solid lignin components may attract and retain trace metals from the leachate.

Several organic components have been detected in landfill leachate added to the leachate components given in Table 3.1. The organic compounds and classes reported by several investigations are respectively presented by Table 3.2 (Sridharan and Didier 1988).

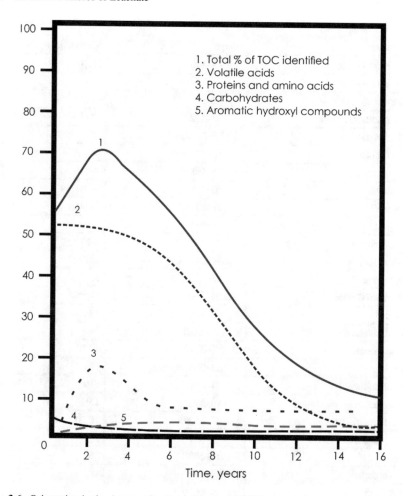

1. Total % of TOC identified
2. Volatile acids
3. Proteins and amino acids
4. Carbohydrates
5. Aromatic hydroxyl compounds

Fig. 3.6 Orientation in the detected fraction of leachate TOC in comparison with landfill age

Table 3.2 Particular organic components detected in landfill leachates

Parameter	Range	Median (μg/l)
Acenaphthene	13.9–21.3	17.60
Benzene	1–1630	11.10
Bis-2-ethylliexylphthalate	91–7900	1050
Butylbenzylphthalate	10–64.1	37.05
Carbontetrachloride	3–995	28.00
Chlorobenzene	3–188	25.20
Chloroethane	2–73,017 μg/l	
Chloroform	4.4–16	7.14

(continued)

Table 3.2 (continued)

Parameter	Range	Median (μg/l)
Di-N-butylphthalene	13–540	28.70
Di-N-octylphthalene	16.1–542	110
Dibromochloromcthane	22–160	91
Dichlorodifluoromethane	100–242.1	171.05
Dichloromethane	27.6–58,200	483
Diethylphthalate	12–230	44
Ethylbenzene	1–1680	43.50
Fluoranthene	9.56–723	39.10
Fluorene	21–32.6	26.80
Fluorotrichloromethane	1–183	34
Formaldehyde	1–1.4	1.20
Halogen,totalorganic	0.0039–33,400	623.50
Lsophorone	3.18–520	76
Methylethylketone	2100–3700	19,550
Naphthalene	4.6–186	33.75
p·dichlorobenzene	8.1–1220	14
Phenanthrene	1.1–2170	50.70
Phenol	0.12–264	174
Phenolics, total	1–232	619
Tanninandlignin, combined	410–1400	1.94
Tetrachloroethylene	1–11,800	16.30
Tetrahydrofuran	1–372.2	730
Toluene	10–3000	360
Trichloroethylene	9.4–2472	19
Vinylchloride	10–3000	230
Xylene	9.4–2472	50

3.4 Carbon, Nitrogen and Phosphorus Proportions at Leachate

Leachate has a great number of organic materials evaluated as biochemical oxygen demand, chemical oxygen demand, ammonia, halogenated hydrocarbons, and heavy metals. Studies have indicated that organic P is moving in the ground, and can introduce a significant ratio of P available in leachate (Worrell and Vesilind 2011).

The ammonia–nitrogen that is present in leachate is elicited from the nitrogen available in the waste; the dosage is changing according to the level of waste solubilization and/or leaching. Yet, the nitrogen level of MSW is below 1%, according to the wet-weight basis, and for the most part, it is made up of the proteins available in

Table 3.3 Ammonia–nitrogen dosage limits for landfills

Stabilization phase	Concentration (mg/l as N)	
	Conventional landfills	Bioreactor landfills
Transition	120–125	76–125
Acid formation	2–1030	0–1800
Methane fermentation	6–430	32–1850
Final maturation	6–430	420–580

homestead wastes, food wastes, and bio-solids. When the proteins are broken down and fermented by microorganisms, ammonia–nitrogen is generated. This reaction is known as ammonification. Researchers introduce dosages that are below detection levels and over 5000 mg/l.

Leachate material is totally changing, relating greatly to waste structure, the moisture level of the waste, and landfill age. Table 3.3 gives limits of ammonia–nitrogen dosage for both common and bioreactor landfills.

At landfills that leachate recirculation is exercised, leachate ammonia dosages can compile at so much greater levels than during common single pass leaching, accordingly, providing a leachate depletion issue. Leachates of bioreactor landfills commonly have ammonia nitrogen content less than 5000 mg/l. This level is about 100 times more than ammonia nitrogen content typically seen in urban wastewater. Such great content of ammonia can introduce many issues to the environment including the excessive richness of nutrients in the surface water. Other destructive influences derived from nitrogenous depletion are decreased chlorine sterilization yield, added dissolved oxygen discharge in input waters, negative impacts on people's health, and less appropriate for using again. Because of the toxic influences that ammonia has, ammonium content must be kept to a favorable amount, <10 mg/l, before it is depleted.

3.5 Microbiological Structure of Leachate

The information available regarding the microbiological structure is rare as opposed to the chemical properties of leachate which have been greatly generated.

Municipal Solid Waste (MSW) has a very huge microbial content and may be polluted by pathogenetic micro-organisms. Wastes usually have animal ordure and corpse, diapers, effluent sludge and every so often hospital waste that is like the machine for pathogenic micro-organisms and therefore regarded as a health danger (Christensen and Kjeldsen 1989; Shayesteh et al. 2020; Pazoki et al. 2018).

The first microbiological researches were conducted on hygienic landfills, examining and focusing on health and waste related issues of micro-organisms, particularly microbes and viruses with ordure origin, available in waste and leachate. Newer researches began to analyze the number and physiological activity of micro-organisms that play role in stabilization of MSW, in more detail, which contains outer cell enzymatic interaction and paced chemical procedures to evaluate the number of methanogenic micro-organisms.

3.5.1 Bacteria

Great leachate–attached bacterial mass with changing quality and structure based upon landfill age- was seen in many types of research. Existence of ordure streptococci and ordure coliforms show respectively ordure pollution and pollution brought about by warm-blooded animals. However, some of the coliforms have been employed to show the possible existence of pathogenic materials.

The existence of bacterial in a sanitary landfill is in turn correlated with the temperature in the landfill. In addition, bacteria grow and exist at temperatures of more than 60 °C. The bacterial movement reduces along with pH level and the collaborative impact of temperature and pH increases the speed of this phenomenon, even more (Andreottola et al. 1990; Lay et al. 1998).

The condensation of fecal index bacteria commonly decreases when landfill gets older and finally attains levels that are no more perceptible. In lysimeter Tests including bog mud and MSW, the existence of streptococci was no more perceived after two years.

3.5.2 Viruses

Unlike viruses are parasitic cells that are impotent to reproduce without the host. The existence of RNA viruses in leachate can be ascribed to the existence of fecal from MSW landfills. But, the existence of RNA viruses was only seen in very scarce cases.

3.5.3 Fungi

According to previous studies, little is known about the existence of Fungi in leachate. A microorganism that lives on dead or decaying organic such as Aspergillus, Penicillium, and Fusarium which are not infective, are the types seen by many researchers. Allescheriaboydii is the only infective fungus detected which can result in Madura foot cyst.

3.5.4 Parasites

There are no figures in the previous studies about the existence of parasites in leachate. Parasites such as protozoa, helminths, and nematodes may be detected in leachate because of the existence of livestock and human faces in landfills and bog muds. Parasitic bladders and seeds are exceedingly invulnerable to fecal micro-organisms. A situation in which de-active infective viruses are typically unprofitable for parasites, especially for nematodes and helminths.

3.6 Factors Affecting Leachate Composition

The chemical compound of leachate hinges on many elements, such as those concern waste pile and position localization and those obtained from landfill layout and operation.

3.6.1 Waste Composition

The decay of waste in landfills, as well as the characteristic of generated leachate, are greatly affected by the quality of waste living portions. Mainly, the existence of materials that are poisonous to bacterial flora may decelerate or interdict biological decay progress with outcomes for the leachate. The proximity of waste to leaching water and also PH and the chemical stability at the solid–liquid joints play a significant role in the organic quantity of the leachate. The bulk of metals are under acidic estate released.

3.6.2 pH

Chemical procedures which are the cornerstone of pile movement in the waste leachate system, including sedimentation, solution, and oxidation and reduction and absorption reactions are affected by pH. It also has an effect on the formation of new species of most of the components in the system. Commonly, acidic estates, which are typical of the acteristic of the first stage of oxygen-less decay of waste, enhance solubilization of chemical components (oxides, hydroxides, and carbonated structures) and reduce the adsorptive quality of waste.

3.6.3 Redox Capacity

Reduction, the second and third stages of oxygen-less decay will impact the solvability of nutrients and metals in leachate.

3.6.4 Landfill Age

Dissimilarity in leachate structure and in the number of contaminants elicited from waste are mostly ascribed to landfill age, interpreted at the time assessed from the first coming of leachate. This element is of utmost significance in the designation of leachate specification controlled by the nature of waste consolidation operation. It ought to be signified that dissimilarity in the structure of leachate does not hinge solely on landfill age but on the level of waste balance and amount of water that permeates into the landfill. Throughout the first years of landfill formation (2–3 years), the contaminant load in leachate typically attains the highest amount and then slowly declines over the coming years. This trend is typically appropriate for organic elements, main indices of organic contamination (COD, BOD, and TOC), and microbiological collection and to principal non-organic ions (heavy metals, Cl, SO_4, etc.).

3.6.5 Waste Age

The impact of waste age on the nature of leachate from a landfill location where recent and grown waste organisms live was examined over a 6-year-term to estimate the influence of waste age on the nature of formed leachate. The findings of the study showed that the leachate nature is under the effect of waste age due to its influence on bacterial development and chemical reaction in the waste pile of landfills (Chian and DeWalle 1977).

Differences appeared in the BOD/COD, COD/TOC, VS/FS, and VFA/TOC proportions of leachate hinge considerably on the age of the landfill. Figure 3.7

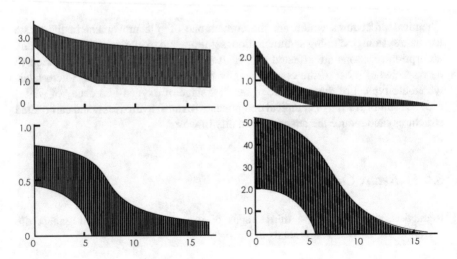

Fig. 3.7 Overtime changes in leachate

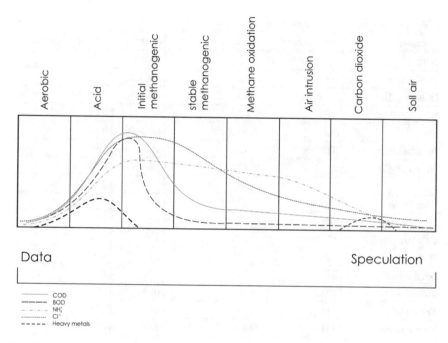

Fig. 3.8 Leachate condition over the lifespan of a landfill

shows the orientation in the type of leachate over the age while Fig. 3.8 indicates the orientation of leachate condition over the life span of a landfill (Chian and DeWalle 1977; Kylefors and Lagerkvist 1997).

Throughout the early years (less than 5), the landfill is in the acidogenic stage and the leachate produced is typically known as young or carbon-based leachate because of the higher dosage of organic carbon. Information regarding the contaminants in the leachate of young waste organisms of the landfill are described in Table 3.4.

Information regarding the contaminants in the leachate of the landfill during a 15-year-period is described in Table 3.5.

3.7 Natural Attenuation of Landfill Leachate

Natural Attenuation (NA), which is also referred to as inherent or inactive remediation, is a method by which normal procedures in soil systems are depended on to decay and disintegrate pollutants. Pollutant dispersal may happen by removal or conversion. If the soil system situation is appropriate for the incidence of NA, it means that the requirement to refine the soil system to very low remaining pollutant concentrations is reduced to the smallest amount (Bjerg et al. 2011; Baun et al. 2003).

Table 3.4 Contaminants in the leachate of young waste

Parameter	Maximum value	Minimum value	Men	Standard deviation
BOD (mg/l)	6350	13.1	2031.62	2754
COD (mg/l)	9600	226	3641.2	3949.71
DOC (mg/l)	3490	90	1240	1421.41
BOD:COD ratio	0.66	0.05	0.31	0.29
pH	6.8	5.92	6.52	0.35
Ammonia (mg/l)	520	103	288	196
Nitrite (mg/l)	2	1	1.8	0.45
Nitrate (mg/l)	2	1	1.8	0.45
TKN (mg/l)	880	162	474	347
Calcium (mg/l)	405	62	181	85
Chloride (mg/l)	1100	275	735	235
Iron (mg/l)	32	2.9	11.24	8.3
Magnesium (mg/l)	215	75	132	37
Sodium (mg/l)	1030	297	688	253
Sulfate (mg/l)	771	7.03	81	175
Phenol (mg/l)	951	1	175	274

Table 3.5 Contaminants in leachate of landfill during a 15-year-period

Parameter	Maximum value	Minimum value	Men	Standard deviation
BOD	870	62	195	202
COD	1510	409	875	256
DOC	650	137	290	167
BOD:COD ratio	0.57	0.05	0.2	0.14
pH	7.38	6.27	5.72	2.64
Ammonia	523	78	260	102
Nitrite	2	0.1	0.72	0.78
Nitrate	2	0.1	0.73	0.77
TKN	663	90	304	118
Magnesium	179	118	153	27.75
Sodium	1370	279	764	450
Sulfate	509	40	196	180
Phenol	1720	3	783	757

Soil systems encompass natural matters such as microbes, aqueous iron, iron minerals, reduced sulfur, and soil organic material. The reciprocal influence of these materials with pollutants can end in natural weakening. The difference between natural weakening procedures and related operations that end in the movement of pollutants from the aqueous stage to the solid stage (adsorption and sedimentation) has been an area of disagreement. these two series of operations are interdepended, but NA indicates a reduction in the dosage of the considered pollutant within the general soil system and not a transition from one stage to another which is much like the adsorption. Both NA and phase-change procedures end in a reduced pollutant level in the aqueous phase. Table 3.6 explains the natural attenuation procedures of organic pollutants (EPA 1995).

Treatment operation for washing out polluted places and keeping groundwater resources safe, like digging soil along with refinement or observed storage, pump-and-treat, or methods to manually maintain contamination, are costly or have commonly low impact.

Fortunately, studies carried out essentially throughout the 1990s indicated that there is a possibility for the biological decay of many organic pollutants. In addition, pollutants downtrend in dosage or defer with regard to groundwater stream by physicochemical procedures as hole water blending and adsorption. The integrated

Table 3.6 Natural attenuation process of organic pollutants

Mechanism	Description	Potential for BTEX attenuation
Biological aerobic	Microbes utilize oxygen as an electron acceptor to convert contaminant to CO_2, water, and biomass	Most significant attenuation mechanism if sufficient oxygen is present. soil air $O_2 > 2\%$, groundwater D.O > 1–2 mg/l
Anaerobic Denitrification Sulfate-reducing Methanogenic Fe-reducing	Alternative electron acceptors (e.g., NO_3^-, SO_2^{4-}, Fe_3^+) are utilized by microbes to degrade contaminants	Rates are typically much slower than for aerobic biodegradation: toluene is the only component of BTEX that has been shown to consistently degrade
Hypoxic	Secondary electron acceptor required at low oxygen content for biodegradation of contaminants	Has not been demonstrated in the field for BTEX
Physical volatilization	Contaminants are removed from groundwater by volatilization to the vapor phase in the unsaturated zone	Normally minor contribution relative to biodegradation. More significant for the shallow or highly fluctuating water tables
Dispersion	Mechanical mixing and molecular diffusion processes reduce concentrations	It decreases concentrations but does not result in a net loss of mass
Sorption	Contaminants partition between the aqueous phase and the soil matrix. Sorption is controlled by the organic carbon content of the soil. Soil mineralogy and grain size	Sorption retards plume migration but doesn't permanently remove BTEX from soil or groundwater as desorption may occur

effect of naturally aroused procedures ending in a diminished distribute of contamination or a drop in dosage away from the origin is known as NA (Murray and Beck 1990; Pazoki et al. 2015b).

Inherent- or inactive bioremediation shows microbiological elements of NA, while forged incitement of biological decay is known as live or strengthened bioremediation. When NA is potent enough it will impede the advanced transition of contaminant and an unvarying condition is raised after a period of time. The depletion of the contamination resource and gradual decline of the downfall may finally result in contracted plumes (Wiedemeier et al. 1999).

Current rules on NA refer to that sites have to be checked promptly to confirm the duration of NA procedures. As a result, the accomplishment of NA is called Monitored NA (MNA). Benefits of MNA in comparison with other remediation methods cover less amount of remediation wastes, diminished human vulnerability, fewer remediation expenses, and preferable assignment of remediation resources. A drawback of MNA is that time setting for remediation are long and not simple to foretell; thus, the long-term track is required. Besides, the validity of the resumption of biodegradation is of critical importance. An intense mount in demand of MNA at the Superfund plan and especially the Underground Storage Tank (UST) plan took place in the USA during the late 1990s. But, errors in data elucidation are actually unavoidable, or the degree of documentation that is being avowed is not enough, ending in the use of MNA in places where its productiveness has not been shown properly. Besides, viewpoints on whether NA is a suitable strategy for controlling groundwater pollution are highly polarized. Furthermore, critics believe that controlled acceptance of MNA occurred too fast.

In general, 14 instruction for use of MNA from a series of the corporation were lately reconsidered by the USA's National Research Council's (NRC's) Committee on Intrinsic Remediation. The main outcomes of this review were:

1. MNA is a fixed remedy for only some varieties of pollutants,
2. Severe rules are required to make sure that NA potential is reviewed thoroughly,
3. MNA ought to be approved as a conventional remedy for pollution exclusively if the procedures are recorded to be functional and are balanced.

Besides, the review showed that the main point of representing the potency of NA at a place is organizing the logical connection between loss of pollutants and the normal procedures accountable for the damage. Yet, the procedures controlling the below surface destination of many pollutants are only in part realized and are yet subjects of ongoing studies.

NA incidence makes aquifers proper as normal remedy systems of strata below the surface. Decay, adsorption, solution, volatilization, sedimentation, and exchange of ions are operations which weakens contamination, but only (microbial) decay actually eliminate the pile of organic pollutant and is necessary for gaining resting plumes.

The decay of outspread organic contamination usually ends in a sequence of oxidation and reduction zones from the origin to the outer part of a plume as available electron-acceptors are employed in a privileged order. Forecasting the development of oxidation and reduction regions is of great significance for the measurement of NA, as the potency and level of decay of certain organic materials hinge on the oxidation and reduction conditions of the undersurface strata, i.e., the existence of oxygen, nitrate, iron oxide, and sulfate. For instance, benzene is properly decayed with oxygen, but proportions are much decelerated or even zero below oxygen-less state and alteration in each place hinges on the existing electron-acceptor.

Microorganisms alter organic components through the electron-transfer process so as to draw free power by combining ATP. Generally, organic combinations act as a reducing agent (e.g. BTEX), but non-organic materials (e.g. O_2, SO_4^{2-}) are used as the oxidizing agent. Yet, some organic combinations (e.g. halogenated hydrocarbons: trichloroethylene, etc.) act as the oxidizing agent through chemical reduction of dehalogenation, and demand (natural) reducing agent in aquifers such as organic material or pyrite. In addition, fermentation alters organic combinations, but external oxidizing agents are required for full oxidation to CO_2. The decay of organic pollutants explains the progress of oxidation and reduction conditions in hydrocarbon plumes. On the other hand, Dissolved Organic Carbon (DOC) in landfill leachate contains primarily humic-fulvic, and fatty acids, but certain organic pollutants give typically below 0.1–1%. Thus, the decay of DOC forces the progress of reduction in a leachate plume, and therefore explains the spatial–temporal decay with a potency of certain organic chemicals. Thermodynamics largely explains the order in which microorganisms use electron-acceptors since microorganisms are willing to carry out oxidation and reduction close to thermodynamic balance, which brings about the noticed sequence of oxidation and reduction conditions (Fig. 3.9).

The Gibbs release energy for oxidizing organic carbon drops at neutral pH in the sequence: O_2, NO_3^-, Mn(IV)-oxide, Fe(III)-oxide, SO_4^{2-} and CO_2 (Fig. 3.10)

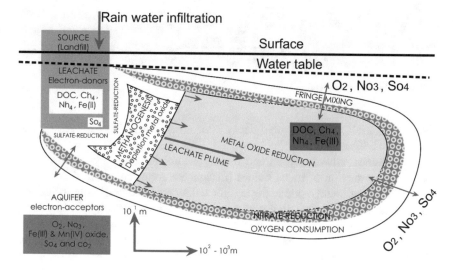

Fig. 3.9 Sequences of redox conditions

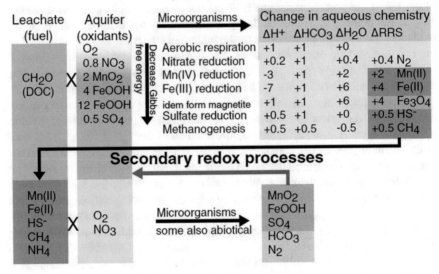

Primary redox processes

Leachate (fuel)	Aquifer (oxidants)	Microorganisms →	Change in aqueous chemistry			
			ΔH^+	ΔHCO_3	ΔH_2O	ΔRRS
	O_2	Aerobic respiration	+1	+1	+0	
	$0.8\ NO_3$	Nitrate reduction	+0.2	+1	+0.4	$+0.4\ N_2$
CH_2O	$2\ MnO_2$	Mn(IV) reduction	-3	+1	+2	+2 Mn(II)
(DOC) X	$4\ FeOOH$	Fe(III) reduction	-7	+1	+6	+4 Fe(II)
	$12\ FeOOH$	idem form magnetite	+1	+1	+6	+4 Fe_3O_4
	$0.5\ SO_4$	Sulfate reduction	+0.5	+1	+0	$+0.5\ HS^-$
		Methanogenesis	+0.5	+0.5	-0.5	$+0.5\ CH_4$

(Decrease Gibbs free energy ↓)

Secondary redox processes

Mn(II)			MnO_2
Fe(II)		Microorganisms →	FeOOH
HS^- X	O_2	some also abiotical	SO_4
CH_4	NO_3		HCO_3
NH_4			N_2

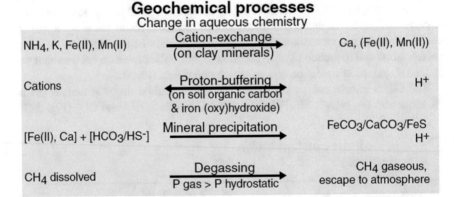

Geochemical processes
Change in aqueous chemistry

	Cation-exchange (on clay minerals) →	Ca, (Fe(II), Mn(II))
NH_4, K, Fe(II), Mn(II)		
Cations	Proton-buffering (on soil organic carbon & iron (oxy)hydroxide) →	H^+
[Fe(II), Ca] + [HCO_3/HS^-]	Mineral precipitation →	$FeCO_3/CaCO_3/FeS$ H^+
CH_4 dissolved	Degassing P gas > P hydrostatic →	CH_4 gaseous, escape to atmosphere

Fig. 3.10 Primary redox reaction

(Lopez et al. 2004). Thus, aerobic decay after nitrate reduction oxidizes organic carbon at the edge of plumes.

In the plumes, where oxygen and nitrate are not present, oxygen-less decay appears, and reduction of metal oxides (manganese followed by iron) continues. A large amount of metal oxide in the aquifer sediments in contrast with the dosage of DOC makes reducing the solution of metal oxide an important oxygen-less decay reaction.

Metal oxides discharged from the origin are accompanied by sulfate reduction and in the end methanogenesis. Sulfate reduction may happen close to the landfill area, since leachate is a source of sulfate, and at the edge, where sulfate from the pristine groundwater blends and metal oxides is discharged (Fig. 3.8).

Fig. 3.11 Synthetic liner on the slope, ready for earth cover

Areas of iron reduction, sulfate reduction, and methanogenesis are usually seen to cover based on pH, redox types' dosage, and solvability of iron-oxide minerals. It can be described by the concurrent incidence of these oxidation and reduction procedures by a partial balance technique, where fermentation of organic material is overall rate restricting, while microbial oxidation is considered to be near the equilibrium. Primary oxidation and reduction procedures include the oxidation of organic carbon and freeing of oxidation and reduction species (CH_4, H_2S, Fe(II), and Mn(II)). These reducing agents become oxidized in secondary oxidation and reduction reaction. At the edge of a leachate plume specifically, secondary oxidation and reduction reaction occur and challenge the organic carbon decay procedure for existing electron-acceptors (Figs. 3.9 and 3.10).

Finally, the decay rate for organic chemicals hinges mainly on the existence of oxidizing/reducing agents in the aquifer, given that the required microorganisms are available. Rates will drop in time when electron-acceptors are used, which leads to the expansion of the plume with the unvarying conditions.

3.8   Leachate Collection

Leachate is headed toward low spots at the base of the landfill visa the use of an effective sewerage layer made up of sand, gravel, or stone material. Pierced pipes are located at in-depth points to gather leachate and are tilted to allow the moisture to exit the landfill (Delarestaghi et al. 2018).

The primary goal of the lineation of a landfill cell is to keep down the potential for groundwater pollution. The liner plays the role of an obstacle between the buried waste and the groundwater and creates a catch dip for leachate built by the landfill. The leachate that is gathered inside the cell has to be removed from above the liner as fast as possible since the RCRA Subtitle D adjustment confine the head of the leachate (free liquid depth) on a liner system to 30 cm.

Leachate is usually eliminated by two methods: gravity flow or pumping. The various elements of a leachate collection system for an MSW landfill usually are (Delarestaghi et al. 2018):

- Protection and drainage layers.
- Perforated collection lateral and header pipes.
- Pumping station sump.
- Leachate pumps.
- Pump controls.
- Pumping station appurtenances.
- Force main or gravity sewer line.

3.9   Environmental Influences of Leachate

3.9.1   Influences on Groundwater

Pollution of groundwater by waste landfill leachate is assumed to be of the main biological issues. Table 2.2 indicates that the most generally recorded evidence of groundwater contamination in recently industrialized countries are rapidly increasing concentrations of the main ions sodium, chloride, and bicarbonate together with ammonium and iron. Yet, it is not clear to what extent insufficient analytical equipment may confine this list.

3.9.2   Consequences of Oxidation and Reduction Changes

The existence or absence of some species can be used as indices of oxidation and reduction conditions. Nitrate normally hints at oxidizing conditions. After being solved, oxygen is used in the degradation of organic matter, nitrate is gone by denitrification to nitrogen gas and recedes from the view. This comes increasingly behind the reduction of insolvable Mn(IV) species to solvable Mn(II), reduction of insolvable Fe(III) to solvable Fe(II), reduction of sulfate to sulfide and in the end, reduction of carbon dioxide to methane.

These oxidation and reduction variation may have important significance for groundwater quality. In Mae Hia, Thailand, a situation in the groundwater plume during the arid season seems to be mostly anaerobic with ammonia available and manganese being generally identified at inappropriate concentrations. In the wet season, the chemistry looks more complicated and nitrate is substituted with ammonia as a result of oxidizing groundwater reload, but manganese continues urging. Table 3.7 describes this plan of oxidation and reduction alterations.

Table 3.7 Summary of chosen emergence of groundwater contamination from waste disposal in recently industrialized countries

Country	Disposal method	Waste type	Groundwater contamination indicators	References
Argentina	Unlined fill with burial	Municipal	CI, HCO_3, CI/HCO_3, ZN	Martinez et al. (1993)
Brazil	Sanitary landfill	Industrial Municipal	Na, Cl, NH_4	Vendrame and Pinho (1997)
Greece	Sanitary landfill	Municipal	SEC, hardness, Cl, P, metal, NH_4, NO_3	Loizidou and Kapetanioos (1993)
India	Open dumping in low lying areas	Municipal	TDS, Cl, Cr, Ni, Cu, CN, NH_4	Olaniya and Saxena (1977)
Romania	Old quarry	Industrial Municipal Medical	Na, Cl, Cr, Ni, Cu, CN, NH_4	Mocanu et al. (1977)
Ukraine	Solid waste	Industrial Municipal	Bioindicators—suppression of microbiological activity	Magmedov and Yakovelva (1977)

3.9.3 Organic Loading

There have been proportionally few researches on the influence of organic combinations leached from urban waste locations to groundwater; it was understood that most of the compounds emanated from disintegrated plant material such as aliphatic and aromatic acids, phenols and terpenes. insignificant components were either chlorinated or non-chlorinated hydrocarbons, nitrogen-containing combinations, alkyl phenol polyethoxylated and alkyl phosphates; used a group of C4 to C7 carboxylic acids with the predominance of even carbon numbered compounds observed in the leachate as indices of organic contamination from leachate.

So rare and particular organic combinations were identified in the current research locations, possibly indicating the principal leachate chemistry.

Identified substances were:

– Plasticizers, generally the phthalate esters, with a single model of dioctyl adipate.
– Diesel or greasy oil possibly from properly pumped equipment and prepared engine oil.
– A separated example of chloroaniline, an industrial chemical.

The non-particular organic charge from leachate, as shown by assessments of solved organic carbon, may create a greater danger because of the formation of trihalomethanes throughout water sterilization with chlorine.

3.10   Solution for Leachate Issues in Landfill

As stated above, groundwater contamination is the most common effect of leachate discharge into the environment. The first step for avoiding this issue is that we should place the landfill away from the groundwater stand or also away from groundwater abstraction borehole. So, more consideration has to be given to learning the hydrogeology of the region so as to detect the best place for the landfill.

Another step is to place the landfill in low penetrable soil or to use sealed liners to cover wastes and leachate.

These days, the leachate control entails not only landfill engineering but the implication of waste management itself. Pile transfer procedure creates leachate contamination. There are 3 physical stages in the reactor:

- The solid phase (waste).
- The liquid phase (leachate).
- The gas phase.

Waste entering the landfill reactor undergoes biological, chemical, and physical changes which are observed among other affective factors, by water input streams.

The liquid stage is enhanced by solubilized or suspended organic ions from the solid phase. There is Carbon in the form of CO_2 and CH_4 in the gas stage. The rationale for discharging of leachate into the environment under more limited visions are:

- Groundwater contamination at landfills.
- The bigger the size of landfill beneath the bigger danger poses.
- Demand to agree with more and more limited rules about quality standards for wastewater depletion.
- The size of residue will be declined with a good technique but more dangerous waste may require to be landfilled.
- The reposition of leachate may be a negative element with regard to geotechnical solidity in both on a downhill or on the ground.
- Leachate control equipment ought to last and their productiveness is guaranteed over the long term.

References

Andreottola G, Cannas P, Cossu R (1990) Overview on landfill leachate quality. Caligary, Italy, CISA Environmental Sanitary Engineering Centre

Baun A, Reitzel LA, Ledin A, Christensen TH, Bjerg PL (2003) Natural attenuation of xenobiotic organic compounds in a landfill leachate plume (Vejen, Denmark). J Contam Hydrol 65(3–4):269–291

Bjerg PL, Tuxen N, Reitzel LA, Albrechtsen HJ, Kjeldsen P (2011) Natural attenuation processes in landfill leachate plumes at three Danish sites. Groundwater 49(5):688–705

Canziani R, Cossu R (2012) 4.1 Landfill hydrology and leachate production. In: Sanitary landfilling: process, technology and environmental impact, p 185

Chian ES, DeWalle FB (1977) Characterization of soluble organic matter in leachate. Environ Sci Technol 11(2):158–163

Christensen TH, Kjeldsen P (1989) Basic biochemical processes in landfills. In: Sanitary landfilling: process, technology, and environmental impact. Academic Press, New York, p 29–49 (9 fig, 3 tab, 34 ref.)

Christensen TH, Cossu R, Stegmann R (eds) (2005) Landfilling of waste: leachate. CRC Press

Delarestaghi RM, Ghasemzadeh R, Mirani M, Yaghoubzadeh P (2018) The comparison between different waste management methods of Tabas city with life cycle assessment assessment. J Environ Sci Stud 3(3):782–793

Ehrig HJ (1989) Leachate quality. In: Sanitary landfilling: process, technology, and environmental impact. Academic Press, New York, p 213–229 (10 fig, 3 tab, 4 ref.)

EPA U (1995) Manual, ground-water and leachate treatment systems. EPA/625/R-94/005, 119

Ghasemzade R, Pazoki M (2017) Estimation and modeling of gas emissions in municipal landfill (Case study: Landfill of Jiroft City). Pollution 3(4):689–700

Kylefors K, Lagerkvist A (1997) Changes of leachate quality with degradation phases and time. In: Proceedings of the sixth international landfill symposium. Cagliari, Italy

Lay JJ, Li YY, Noike T (1998) Developments of bacterial population and methanogenic activity in a laboratory-scale landfill bioreactor. Water Res 32(12):3673–3679

Lopez A, Pagano M, Volpe A, Di Pinto AC (2004) Fenton's pre-treatment of mature landfill leachate. Chemosphere 54(7):1005–1010

Murray HE, Beck JN (1990) Concentrations of synthetic organic chemicals in leachate from a municipal landfill. Environ Pollut 67(3):195–203

Pazoki M, Delarestaghi RM, Rezvanian MR, Ghasemzade R, Dalaei P (2015a) Gas production potential in the landfill of Tehran by landfill methane outreach program. Jundishapur J Health Sci 7(4)

Pazoki M, Pari MA, Dalaei P, Ghasemzadeh R (2015b) Environmental impact assessment of a water transfer project. Jundishapur J Health Sci 7(3)

Pazoki M, Abdoli MA, Ghasemzade R, Dalaei P, Ahmadi Pari M (2016) Comparative evaluation of poly urethane and poly vinyl chloride in lining concrete sewer pipes for preventing biological corrosion. Int J Environ Res 10(2):305–312

Pazoki M, Ghasemzade R, Ziaee P (2017) Simulation of municipal landfill leachate movement in soil by HYDRUS-1D model. Adv Environ Technol 3(3):177–184

Pazoki M, Ghasemzadeh R, Yavari M, Abdoli M (2018) Analysis of photocatalyst degradation of erythromycin with titanium dioxide nanoparticle modified by silver. Nashrieh Shimi va Mohandesi Shimi Iran 37(1):63–72

Qasim SR, Chiang W (1994) Sanitary landfill leachate: generation, control and treatment. CRC Press

Shayesteh AA, Kooshekan O, Khadivpour F, Kian M, Ghasemzadeh R, Pazoki M (2020) Industrial waste management using the rapid impact assessment matrix method for an industrial park. Global J Environ Sci Manage 6(2):261–274

Sridharan L, Didier P (1988) Leachate quality from containment landfills in Wisconsin. In: Proceedings of the 5th international solid waste conference. International Solid Waste Management Association, Silver Spring, MD

Tchobanoglous G, Theisen H, Vigil S (1993) Integrated solid waste management: engineering principles and management issues. McGraw-Hill.

Wiedemeier TH, Rifai HS, Newell CJ, Wilson JT (1999) Natural attenuation of fuels and chlorinated solvents in the subsurface. Wiley

Worrell WA, Vesilind PA (2011) Solid waste engineering. Cengage Learning

Chapter 4
Leachate Quantity

In the foregoing chapter, the leachate characteristic was analyzed. In the current chapter, the main focus is on leachate quantity. However, the principal goal is to deliver proper patterns for the quantitative assessment of leachate.

4.1 Introduction

One of the most significant topics about placing, projecting, plan, function, and long term handling of an urban solid waste landfill is controlling the it's leachate.

Commonly, the quantity of leachate is a straightforward concern of the amount of external water entering the landfill (Pazoki et al. 2017; Tchobanoglous et al. 1993a). Figure 4.1 introduces the various water outlay and its motion in the landfill leading to leachate generation.

Until now it has been a prevailing issue to slow down leachate creation due to its potential to water contamination. However, leachate is regarded more and more like the way by which the contamination potent of wastes may be freed in a monitored manner. Yet, leachate is more and more regarded as the ways freeing wastes with contamination capacity are handled. Albeit, slowing down leachate formation is still profoundly inserted in more national and EU regulation (such as the recent Landfill Directive), certification of the fact that water is demanded as a response and means of conveying is outspread. Upcoming techniques thus presumably concentrate on handling the total leaching procedure, instead of actually reducing leachate quantity, and these techniques may from time to time require steps to add to the quantity of leachate. The capacity to forecast, construe and detect leachate quantity and levels inside landfills will remain serious.

Intentions to improve leachate quantity have typically been planned to reduce them. However, in some places, non-dangerous liquid wastes have been intentionally added to MSW landfills only to cater sufficient humidity to promote decay, and

© Springer Nature Switzerland AG 2020
M. Pazoki and R. Ghasemzadeh, *Municipal Landfill Leachate Management*,
Environmental Science and Engineering,
https://doi.org/10.1007/978-3-030-50212-6_4

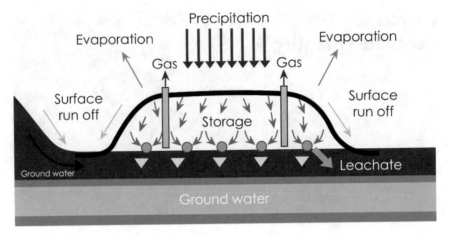

Fig. 4.1 Water transition in the landfill

some leachate handling techniques may demand added rates of leaching so as to attain terminal storage quantity in a logical period of time (e.g. 30 to 50 years). Amendment proceedings are (Tchobanoglous et al. 1993b):

1. Site position, in order to keep away from groundwater depletion region and slow down leachate creation and placement in groundwater depletion region to slow down the danger of groundwater contamination.
2. Site engineering with liners, cut-off hedge, surface water deflection, and low-penetrance upper cover and,
3. Acting in separated cells to limit the region of waste lay bare to precipitation.

The Best Available Technology (BAT) for management of leachate quantity is totally related to the acting strategy. For covering, it is critical to evade or limit leachate. For a flushing bioreactor or a non-organic leaching landfill, leaching rates must be larger than the calculated rate in many locations and would need to be unnaturally elevated, for example by recirculation (flushing bioreactor) or maximization of the infiltration rate/inclusion of water (primary stages of non-organic leaching landfill).

4.2   The Exigence of Estimating Leachate Production Rate

Awareness of the leachate quantity is significant largely for (Lu et al. 1985; Pazoki et al. 2016):

– Use of the proper leachate handling method
– Projection of leachate collection systems
– Projection of leachate remedial equipment

- Specifying admissibility of offsite remedy
- Measure offsite movement potential
- Measure the contamination potential

Use of proper leachate handling method: For the dry and semi-dry region, the quantity of leachate produced may be disregarded and may have no inadmissible influence on the environment; thus, leachate operation may be confined. For rainy regions, where considerable amounts of leachate may be produced, controlled and leachate remedy may be demanded as a short-term leachate handling method.

Projection of leachate collection systems: For collection methods to be favored, a sealed soil hedge or forged liner must be regarded to confine the leachate, and to transport it to the disposal location. The most regular type of collection system employs gravity sewerage and contains a layer of sand and/or gravel under-covered with pierced tubes that transport the leachate to a collection spot. Accordingly, being aware of the leachate quantity is a requirement for specifying tube interval and their diameter. In case of creating open canals for carrying leachate into remedy equipment, we also need figures about leachate quantity to plan canals.

Projection of leachate remedial equipment/disposal regions: Quantitative analysis of leachate ought to be created before determining which remedial/disposal method is to be used. Where considerable amounts of leachate are produced, it may be pivotal to confine the regions subjected to direct sedimentation, as these regions may generate large shifts in leachate quantity. The establishment of small cells may be reflected in to reduce leachate production. The leachate collection system should be intended to merge average climax leachate amounts.

Measure the contamination capacity: as a function of flux.

Measure the capacity for recirculation of leachate: In some weather conditions with significant downfall throughout humid seasons, recirculation of leachate may be confined to decrease the risk of landfill tilt and "overflow" of leachate to the neighboring environment.

Measure the hydraulic stress on the planned liner system: to measure the intensity of leachate freed to the environment.

4.3   Elements Affecting Leachate Quantity

The proportion of leachate production is influenced by a number of elements, including (Lu et al. 1985):

- Waste characteristic, structure (water level) and compressed congestion
- Weather condition (including precipitation)
- Average annual temperature
- Mapping
- Landfill cover
- Plant cover
- Groundwater impacts.

– Cell size and gradual processing of the disposal region

   Functional methods exercised at the landfill.

4.3.1 Waste Type, Structure (Water Level) and Compressed Congestion

Kind of a waste, the water level of the waste and its shape (volume, shredded, etc.) influence both the quantity and quality of the leachate. In addition, leachate generation could be larger with less compression which will decrease the percolation rate.

4.3.2 Climate Condition Elements

Leachate quantity is affected by the level of rainfall (e.g., rain, snow, etc.) on the landfill location, the external runoff, evapotranspiration, and percolation or intervention of groundwater draining through landfill. All these elements are classified in the weather element which showed huge influence on the generation of leachate.

4.3.2.1 Average Annual Temperature

The higher the average annual temperature is, the more considerable the quantity of rainfall that will dry up and not penetrate into waste so decreasing the quantity of leachate production. Local climate condition is a great issue and is best explained as two choices including a dry season (no downfall for a maximum of 5 months) or a wet season with extreme downfall events. The impact of the weather condition on leachate generation is intricate: in proportionally warm weather, for instance, the addition of leachate generation after rainfall is commonly fast.

Moreover, leachate is produced mainly from downfall and then is largely affected by weather conditions such as downfall and evaporation.

In dry weather, almost no overplus leachate takes place; in semi-dry regions, leachate may be produced irregularly or only at specified times of the years. In humid weather conditions, landfills may generate considerable amounts of leachate throughout a year.

4.3.3   Topography

Mapping influences the hurricane water "run-on" and "run-off" from the location and therefore the level of water coming in and out of the location. The surrounding hurricane water channel should be structured to put surface water run-on off from the location and the landfill cover structured to foster run-off and decrease percolation.

4.3.4   Landfill Cover

Throughout the early 1990's considerable attention was concentrated on reducing to the lowest amount of leachate through the upper cover of the landfill. Growth of less penetrable covers, upper sewerage systems, capillary obstacles and promoted evapotranspiration by the selection of plant cover were total techniques reviewed to enhance the reduction of leachate quantity.

Among the techniques employed to decrease the generation of leachate and, therefore, hydraulic ends producing flow from a barred landfill is to set a limit of low penetrable material (e.g. clay or high compression polyethylene- HDPE) over the waste deposit in order to decrease percolation of rainfall. These should be documented in condition measurement because if a landfill is limited to prevent rainwater influx, decreasing leachate quantity, more focused leachate will be produced. Also, microbiological and biochemical procedures will be hindered and thus lengthening the decay phase and the activity of the waste may be for decades or even centuries. Groundwater contamination capacity from older limited landfills may thus be greater than more recent, open landfills (Landreth and Carson 1991).

Due to the main role that conclusive cover acted in decreasing leachate production, it can be classified into two stages.

1. Pre-closure Rate
2. Post-closure rate.

In addition, ordinary and middle cover and the entirety of the final cover and the bottom liner are also regarded as leachate production elements.

4.3.5   Plant Cover

The plant cover has a major role in governing leachate production. It confines percolation by interrupting downfall (with further progressed evaporation from the surface) and by absorbing soil humidity and crossing it back to the atmosphere by depletion. Yet, care must be paid that the bases of the plant cover do not interpenetrate the cap and provide a passage for percolation. Irrigation may be required to guarantee the formation and conservation of plant cover. The plan and quantity of irrigation water used should be of a kind that the evapotranspiration of water by the plant cover will commonly be larger than the amount of irrigation water (Bagchi 1994).

4.3.6   Groundwater Impacts

Groundwater intervention will enhance the leachate quantity and it would occur if the landfill location is built below the groundwater table. Thus, it is demanded in several countries to place beyond the highest groundwater-surface level in order to avoid the intervention of groundwater, and likewise, actions are often required to stay away from the infiltration of surface water. But, sometimes biological conservation assumptions based on setting the bottom of the landfill beneath the groundwater perimeter and thereby producing an internally addressed hydraulic tilt, has been recommended.

Where feasible, as there is inside current to the hydraulic sealed liner, it may be probable to layout the system in a way that engineered hydraulic trap is totally inactive. There are 3 agents that have to be regarded in modeling this engineered hydraulic trap (Bagchi 1994; Pazoki et al. 2015):

1. The top cover in the hydraulic control layer must be handled in a way that "blow out" of either liner does not happen throughout or after setting up.
2. The amount of water gathered by the "hydraulic trap "have to be controllable and the hydrogeological system have to have the volume to cater the water needed to keep the hydraulic trap (if not so, then the top cover in the aquifer will decline and the success level of the trap may be weakened with time).
3. Even if there is a hydraulic trap, some external dissemination of pollutants is to look forward to in many cases. Pollutant transmission analyses are also needed to measure what (if any) influence may appear under these circumstances.

4.4   Leachate Flow Into Soil and into Landfill and Their Variation

Pursuant to the procedure of leachate formation in hygienic landfills, it begins when the quantity of water trespass the current volume of humidity content in solid wastes (field capacity).

In penetrated landfills above an aquifer, water infiltrates to landfills and wastes dumps often amassed or 'piled up' inside or under the landfill. This is because of the production of leachate by decay procedures acting in the waste, added to the rainwater permeating into the waste. The enhanced hydraulic top cover extended promotes the descending and ascending flow of leachate from the landfill or site. Descending flow from the landfill menaces beneath groundwater resources while ascending flow can end in leachate springs yielding water of a poor, often hazardous quality at the perimeter of the waste sediment. Perception of leachate springs or poor

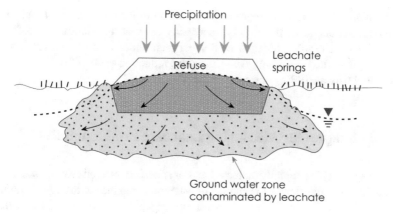

Precipitation

Leachate
springs

Refuse

Ground water zone
contaminated by leachate

Fig. 4.2 Perceptual image of leachate movement from a landfill

water characteristics in neighboring wholes indicates that leachate is being created and is active. Leachate springs introduce considerable risk to public health, so their identification in condition measurement is serious so as to stop entry to these springs (Freeze and Cherry 1979) (Fig. 4.2).

Leachate movement is influenced by the system of waste sedimentation. Compression of waste before sedimentation decreases its penetrability, while common use of the upper cover of soil between the discharges of waste to landfills persuades layer formation. These features necessarily lead to outstanding flow routes through landfills. It was found, for example, that habitats duration for downfall reaching a landfill differed from several days to years. This is cast back in the often provisional character of leachate "springs", which can emerge in rainy seasons but then vanish in arid seasons to leave pieces of soil with changed color. In pieces of potential leachate generation, emphasis should, thus, be on terms towards the end of rainy seasons or after extreme rainy days. To add more, situation measurement requires to consider suspicions in both the forecast and observation of leachate movement from landfills and holes, as a result of complex hydrogeology of waste sediments (Bagchi 1994; Freeze and Cherry 1979).

In spite of the complexity of leachate movement through landfills, substantial facets of underlying pollutant transport can actually be used for the transferring of leachate-derived pollutants from a landfill or waste dump. These comprise the depth of the un-soakable region, the penetrability and humidity level of the ground matters inside the un-soaked region, and the hydraulic flux and surrounding hydraulic tilt of earth-related sections in the soaked region. Poorly leading units beneath the landfill or wasting dump, e.g. material with high clay content or even existence of established forged liner hinder leachate movement. On the contrary, gaps such as seams and hinges in the underlying layer eorfaultsor holes in a liner considerably enhance leachate flux. To measure condition, reaching hydrogeological data and information on the layout and status of potentially established lining mechanisms from both under and downstream of landfills is fundamental. As serious as underlying magnitude and orientation of leachate stream is the detection of the considerable biochemical variations that occur, as powerfully decreasing leachate (oxidation and reduction probability <-100 mV), blended with superficial underlying groundwater, which is soft to powerfully oxidizing (oxidation and reduction probability $> +100$ mV) (Qasim and Chiang 1994; Ghasemzade and Pazoki 2017).

4.5   Techniques for Measuring Leachate Production Rate

Leachate quantity is generally patterned and/or specified through an easy water equilibrium method, considering the amount of water reaching the landfill (that is rainfall, waste moisture in surplus of humidity level capacity of the waste and extra water level such as water in wastewater treatment plant sludge's if permissible) and the level of water departing from the landfill (that is water used in biochemical processes and evaporation) (Bagchi 2004).

Estimating leachate creation can be performed by water balance patterns. There are several patterns, from the very developed and intricate, like the U.S Environmental Protection Agency's (EPA) HELP model with high requirement for information, to the easiest type (Bagchi 2004; Shariatmadari et al. 2010):

$$L = R - Ea.$$

where:

L is the volume of leachate.

R is the volume of rainfall.

Ea is the volume of real evapotranspiration (or simpler evaporation from the earth level).

Both information regarding rain and drying up are gathered simply by weather stations and are often accessible.

Any pattern employed to measure leachate quantity should take into account calculation of the highest value per day, mean quantity per month per year, and mean quantity in a year. Since the amount of created leachate changed considerably from open to closed cells, the measurement in leachate quantity during the lifetime (i.e., from the initial cell to the last cover of the entire disposal area) of the landfill ought to be cast back in leachate quantity evaluation. Water equilibrium patterns are usually exposed to great doubts, given the demand to forecast or measure some of the elements in the equation.

Any water balance pattern should just be used to show the intensity of leachate production. Therefore, for this goal, a relatively easy pattern may demonstrate merely as beneficial as the more intricate patterns.

Water equilibrium calculations are used to forecast leachate quantity at recent landfills and to construe levels and flows at current landfills. They are commonly favorable for the goal of measuring disposal equipment if a margin of error is admissible. They are not usually proper for measuring penetration rates from the foundation of lined landfills since even a considerable penetration amount is typically little in contrast with the general quantity of leachate. The significance and expense of leachate remedies are so that common (e.g. yearly) that re-measurement of water equilibrium is advantageous.

The water equilibrium measurement compares the amounts of all liquids coming in and out of the landfill throughout a given term. Any advance in storage may be available both as attracted or free leachate and this hinges on the storage nature of the waste, the specification of which is vague and intricate. Numerous inlets and outlets of the water equilibrium equation should be regarded but many are commonly disregardable. Effective precipitation (added rain or surplus downfall) is commonly the main input and leachate eliminated for disposal is commonly the main product of pollutant landfills. The obtained quantity of leachate are decreased by two main elements:

1. attraction by the wastes, especially throughout the functional stage; and,
2. Run-off and lateral drainage from finished zones with low-penetrable head.

Humid included in landfill gas and the humid used during oxygen-less fermentation are possibly ignorable except probably at MSW landfills is so arid zones where waste decay may be humid restricted (Delarestaghi et al. 2018; Pazoki et al. 2015).

Every main element of water equilibrium is exposed to errors in measurement, which may usually be completely huge. Some are mechanical, such as natural mistakes in the manner of measuring beneficial downfall, and some are the result of intricacy and expense of catching precise site-dependent information, e.g. for the absorptive quantity of the solid wastes. The net impact of these mistakes hinges on the place and waste input rates. In the less warm wetter zone of the EU, where downfall is much larger in quantity than the possibility of evapotranspiration and Estimated Rate (ER) may be on the sequence of 800–1000 mm/the combined mistakes may result in no greater than a 30% doubt in ER. If the waste input level (and thus the absorptive quantity) is down, then this would as well be the amount of the doubt in leachate quantity. On the other side, in higher temperatures in more arid zones of the EU the actual quantity of the ER could simply be half or twice the probable value. If input rates of absorptive wastes such as MSW were again high then it would be very hard to make an acceptable forecast of leachate creation (Hjelmar et al. 2000; Shayesteh et al. 2020).

Real leachate quantity have been documented for landfills in many areas of the EU and are indicated in table 1, below. Totally, they prove the prediction from the added downfall distribution of Europe that quantity is considerably less in more arid zones, with early an element of ten between the ends (Delarestaghi et al. 2018).

4.5.1 Example of Moisture Mass Equilibrium Measurement for Bioreactor Landfills

The Water Balance Method carried out many per month measurements to calculate the mean moisture level of the waste. It was indeed projected to calculate evapotranspiration from soils and was after that obtained for landfill situation. The Water Balance Method contains a two-tiered method. Method A is an easy equation that only employs elements that most considerably influence the mean humidity level of the waste pile. The simplified equation considers as well that all downfalls exactly on the landfill area will turn into humidity in the waste pile. The major elements that are regarded in the plain equation are:

1. Incoming waste humidity,
2. Downfall (only rain that falls straightly on the landfill's area; believing that all surface runoff from neighboring areas is deflected around the landfill surface),
3. Surplus liquids (recirculated leachate, water, etc.), and
4. Leachate generation.

If landfill owners/operators favor the results of the Method A equation, so no other computation is required. Yet, if another analysis is needed, so landfill owners/operators can continue to Method B which contains a more complex series of measurements. This more intricate way requires many elements such as those referred to in the simplified equation of Method A plus the four factors below:

1. Humid preserved in the landfill surface or cover material,
2. Surface runoff,
3. Surface evaporation, and
4. Evapotranspiration.

Method A:

The potential humidity level of the waste pile in the bioreactor landfill may be measured by a simplified equation of the Water Balance Method as follows:

$$PMC = \frac{(L_0 \times M) + P + LA - LCH}{M + P + LA - LCH} \times 100$$

where,

PMC is measured potential moisture content of the waste pile (% moisture content with regard to weight);

L_o is moisture penetrating with the waste pile (kg moisture/kg total waste pile as received);

M is the whole waste mass in bioreactor cell with received basis (kg whole waste mass as obtained);

P is total precipitation (kg total rainfall);

LA is total liquids added to the waste pile, including recirculated leachate (kg total liquids); and.

LCH is the whole gathered leachate (kg the whole leachate).

If the bioreactor landfill has been at stable condition (i.e., there has been no ups and downs in any of the parameters above) since the bioreactor cell or the whole bioreactor landfill opened, then mean values of *M, P, LA,* and *LCH* can be measured per month instead of totals. However, this assumption is not possible.

By Eq. 1, landfill owners/operators must preserve information and theories employed to specify the values of *Lo, M, P, LA,* and *LCH* for their bioreactor landfill. The following parameters introduce probable instruction for specifying and recording these values.

Lo: Based on *Integrated Solid Waste Management: Engineering Principles and Management Issues*, most MSW in the United States have a humidity level of 15 to 40%, with 25% as usual. The humidity level of MSW hinges mainly on the structure of the waste, the time of the year, and the humidity and climatic conditions of the neighboring environment. For instance, the moisture level of 100 kg of incoming wet waste can be measured as: [(100 kg–*d*)/100 kg],

Here, *d* is the total dry weight (kg) of the solid waste material within the 100 kg of wet waste obtained.

M: To measure total waste mass, waste admission or waste location information is required and should be recorded.

P: Total precipitation per inches of water can be calculated using rainfall over the landfill or from its surrounding climate station information. Change the rainfall data from inches to kilograms of humidity by below equation:

Total precipitation $(P) = $ (in. of total precipitation) * (1 ft/12 in) * (ft2 of bioreactor landfill surface) * (1 gal/0.134 ft3) * (3.78 kg/gal water).

LA: The total amount of increased liquids can be measured by recently obtained at the bioreactor location for planning and functional purposes. For instance, if a closed circle bioreactor with horizontal trenches employs a flow meter to calculate the amount of recirculated leachate, flow meter data can measure total leachate quantity extension (e.g., changing the flow rate per month to kilograms of leachate and then taking sum of the monthly data to calculate a total amount of added liquids).

Water poured at the surface of the landfill by truck could be estimated easily by quantity movement calculation, for example: (gallons of water saved in per tank) * (number of tanks poured on the landfill surface) * (3.78 kg per gallon of water). The kind of liquid addition technique changes based on bioreactor landfill location, thus, the kinds of analysis will change, as well. We suggest that every landfill owner/operator estimate total liquids by methods that are most proper for their bioreactor design.

LCH: Similar to liquids addition, the total amount of leachate generated can be measured by leachate collection data created at the landfill bioreactor for planning, functional, and maybe organization objectives. For instance, if a bioreactor landfill employs a flow meter to calculate the quantity of leachate generated or obtained, then flow meter reading data can be employed to measure general leachate creation (e.g., transforming the total flow rate per month to Kg of leachate and then adding up the monthly data to find a total leachate quantity). The leachate quantity included in Eq. 1 ought to add leachate that is recirculated and any surplus leachate that may be remedied or disintegrated by other methods. We suggest that each landfill proprietor/operator compute total leachate created by procedures most suitable for their own leachate body system plan.

Method B: Developed Series of Measurements

The following elements are demanded inputs for Method B of the Water Balance Method measurements:

- Mean temperatures in degrees Fahrenheit per month (°F)
- Location latitude
- Mean rainfall in inches of water per month
- Landfill surface situation
- Soil & plant cover type for final cover (if any)

The 17 measurement levels of the developed Water Balance Method process are given below.

Levels 1 to 16 of this order measure and prove the infiltration of rainfall into the bioreactor landfill regarding humidity included in the landfill surface or ultimate cover, surface runoff, evaporation wastes, and evapotranspiration. Step 17 is so much like Eq. 1 in Method A. The single distinction between Step 17 and Eq. 1 is that Step 17 substitutes the amount of rainfall with the amount of humidity that infiltrates into the waste pile.

4.5.2   Layer Models

The flow pattern is based on Darcy's law.

$$q = ki^*A.$$

Here, k represents the hydraulic conductivity at places where the flow has been measured; i represents hydraulic tilt at the same spot; q shows the flow at the desired place; A indicates the cross-sectional surface via which flow is happening at the desired place.

4.6 A Computer-Based Method for Estimating Leachate Generation

So many studies have conducted to extend patterns for forecasting leachate quantity from landfills. The model most often employed is the Hydrologic Evaluation of Landfill Performance (HELP). HELP is good for the long-run forecast of leachate quantity and contrast of different scheme remedies. Added to this, Hatfield and Miller (1994) proposed two patterns to better simulate leachate creation at active landfills (Hatfield and Miller 1994): the Deterministic Multiple Linear Reservoir Model (DMLRM) and the Stochastic Multiple Linear Reservoir Model (SMLRM). Several computer programs for estimating leachate generation have been developed, for example, Hydrology Evaluation Leachate Performance (HELP), FULFILL, and SOILINER. These patterns were all based on the namely Water Balance Method (WBM) proposed by the U.S. Environment Protection Agency. Other projects introduce 1D scalar solutions by employing finite margin tools. Every program has essential benefits irrespective of its overall constraints. These constraints are:

(1) Only one level is considered for the measurement without thinking of changes created by the matters on top, or by the solid waste when the depth or height of the landfill is grown;

(2) The reality that cells are not created altogether, or at the same month of the year and since many disposal locations that want to use the available space at the location, cannot close the cell every day with the introduced layer of soil.

(3) The interaction between cells enforced by the structure of nearby cells to yield strips, and/or the structure of other cells on top to make layers is not considered;

(4) These patterns cannot assume the space and time release of leachate creation at the landfill throughout the operation and after cell closure with the layer of soil.

4.6.1 HELP

The HELP pattern is a quasi-2D hydrologic module for performing water equilibrium analyses of landfills, cover systems, and other solid waste pollutant equipment. This pattern allows climate, soil and plan data and employs solution methods that consider the impacts of surface storage, snowmelt, runoff, percolation, evapotranspiration, plant growth, soil humidity storage, lateral underlying sewerage, leachate recirculation, un-soaked vertical sewerage, and penetration to soil, geo-film or composite liners. Landfill systems including various combinations of vegetation, cover soils, waste cells, lateral drain layers, low permeability barrier soils, and synthetic geomembrane liners may be modeled. This pattern assists paced calculation of the quantity of runoff, evapotranspiration, sewerage, leachate collection and liner penetration that may be looked for to result from the procedure of a different kind of landfill plans. The initial goal of the pattern is to aid in the contrast of design remedies (Jang et al. 2002).

4.6.2   SOILINER

SOILINER is a confined-margin estimation of the extremely nonlinear, ruling equation for 1D un-soaked flow in the vertical dimension. SOILINER was created to assume the dynamics of a percolation event among compressed soil liner system incongruity and the association of liner features on the degree of saturation, SOILINER can precisely show percolation for a different type of soil (clay) liner schemes. Considerable characteristics of SOILINER pattern include the potential to assume (Kamaruddin et al. 2017):

1. Multilayered mechanisms
2. Changing initial humidity
3. Changing the situation on the margins of the compressed soil liner flow area.

Considering these properties, SOILINER acts as a thorough device for the planning of liner forms, particularly liner conductivity and depth.

4.7   Discussion on Methods, Their Records, and Outcomes

Early researches contained a grasp of scientific basics of the formation, and chemical structure of leachate (Pazoki et al. 2018; Ghasemzadeh et al. 2017). These were traced by the simulation of the volume measurement of leachate through the Water Balance Method (WBM) and confirming the results with leachate quantity obtained from the basic drains of the landfill (Fenn et al. 1975; Farquhar 1989; Bengtsson et al. 1994). Due to the complicated character of waste, the waste fill has behaved as a being as the modeling was performed on closed landfills. Yet, the difference in the observed calculations and patterned data throughout these periods showed that leachate cannot be logically forecasted by the basic WBM equation. Maybe the most remarkable previous effort to replicate the waste size created in landfills was the expansion of the HELP computer-based pattern. Whilst, the HELP model was found to logically assume the quantity of leachate created from realized waste landfills, its impotence to assume the leachate created during the active course of waste infilling was regarded as a drawback.

More attempts involved applying macro-modeling which bears experimental waste models and customary soil patterns in line with the Hydrologic Evaluation of Landfill Performance (HELP) computer-based pattern to model the measurement of water volume in an emplaced urban solid waste fill throughout both functioning and post-closure courses. This pattern, albeit plain and useful, could not repeat the same leachate quantity but logical average quantity. A lot of attempts have also been made in the application of micro-modeling for replicating the features of emplaced waste at landfill locations; however, there has not been any pattern that can generally and precisely forecast the site quantification. Most of the earlier patterning attempts have been based upon the large quantity and biochemical features of the leachate.

When the pattern used for measuring leachate quantity and characteristics in solid waste landfills were analyzed, it was understood that; while some patterns for replicating leachate quantity have been somehow promising, there is yet a pattern that can rationally simulate the leachate quantity in waste fills.

Added to the above-said models, there has been scalar patterning of various factors of the landfill system. For example, reasonable outcomes were obtained from numerical modeling of gas flow and heat in landfills. Likewise, relatively just findings have been obtained from the scalar simulation of pollutant movement from the landfill and the flow in the seam of a composite basic liner.

Since the main issue in the water leachate from waste landfills is not solely the size but the dosage of pollutants in the sewage stream from the lowest layer, researchers have recently tried studying the movement of solute in the shape of the waste fill. These researches have been somehow promising in using remarkable soil equations to measure the solute pile flux in the bulk waste fills under a stable state. The basic theories connected with has developed from using heat transfer and scattering concept, moveable-immovable concept, double porousness concept to an advanced multiporousness, multi-qualitative, and dual/multi penetrability concept. These patterns have been worked out using definite and accidental methods.

So far, there has not been any attempt to pursue the mass flow of pollutants within the different vertical layers or segments of a waste fill, which is exposed to a recycling flow model. As is seen in the review of researchers on the leached water from landfills above, it is somehow actually infeasible to repeat waste quality fully the same because of the intricate quality of waste. Thus, a "rational result" regarding waste research can be explained as a condition when the volume-related result and the quality orientation of the waste feature are obtained within a sensible degree of accord with the actual calculations.

With the manifestation of computer technology, it is clear that computer-based simulation is commonly applied to make it easier and have a better grasp of complicated systems in relation to time and volume where logical methods may not be simply feasible due to meta-stable or intricate interdepended situations of the system. Mostly, the strictest aspect of scalar patterning is the formulation and measurement of the model. Usually, the most probable method is used to apply simple ones free of obscurity and difficulty, yet potent to logically assume the facts by the real measured information as input. As recirculation of leachate is commonly used to pace the biochemical operations in the landfill to obtain early consolidation, scalar simulation has been performed on the mass flux of a recycling experiment for minor components transport in large-scale experiments. This then allowed actual information to be applied for the credibility and accuracy of the model.

References

Bagchi A (1994) Design, construction, and monitoring of landfills

Bagchi A (2004) Design of landfills and integrated solid waste management. John Wiley & Sons

Bengtsson L, Bendz D, Hogland W, Rosqvist H, Åkesson M (1994) Water balance for landfills of different age. J Hydrol 158(3–4):203-217

Delarestaghi RM, Ghasemzadeh R, Mirani M, Yaghoubzadeh P (2018) The comparison between different waste management methods of Tabas city with life cycle assessment assessment. J Env Sci Stud 3(3):782–793

Farquhar GJ (1989) Leachate: production and characterization. Canadian J Civil Eng 16(3):317–325

Fenn DG, Hanley KJ, DeGeare TV (1975) Use of the water-balance method for predicting leachate generation from solid-waste-disposal sites (No. PB-87-194643/XAB; EPA/SW-168). Environmental Protection Agency, Washington, DC (USA). Office of Solid Waste

Freeze RA, Cherry JA (1979) Groundwater prentice-hall. Englewood Cliffs, NJ, 176:161–177

Ghasemzade R, Pazoki M (2017) Estimation and modeling of gas emissions in municipal landfill (Case study: Landfill of Jiroft City). Pollution 3(4):689–700

Ghasemzadeh R, Pazoki M, Hoveidi H, Heydari R (2017) Effect of temperature on hydrothermal gasification of paper mill waste, case study: the paper mill in North of Iran. J Env Stud 43(1):59–71

Hatfield K, Miller W (1994) Hydrologic management models for operating sanitary landfills. Florida Center for solid and hazardous waste management. Available at: http://www.hinkleycenter.org/landfill-linersystem.html

Hjelmar O, Hansen JB, Andersen L (2000) Leachate emissions from landfills. AFN, Naturvårdsverket

Jang YS, Kim YW, Lee SI (2002). Hydraulic properties and leachate level analysis of Kimpo metropolitan landfill, Korea. Waste Manag 22(3):261–267

Kamaruddin MA, Yusoff MS, Rui LM, Isa AM, Zawawi MH, Alrozi R (2017) An overview of municipal solid waste management and landfill leachate treatment: Malaysia and Asian perspectives. Env Sci Poll Res 24(35):26988–27020

Landreth RE, Carson DA (1991) RCRA cover systems for waste management facilities. In Landfill Closures (pp. 1–9). Elsevier

Lu JC, Eichenberger B, Stearns RJ (1985) Leachate from municipal landfills: production and management

Pazoki M, Abdoli MA, Ghasemzade R, Dalaei P, Ahmadi Pari M (2016) Comparative evaluation of poly urethane and poly vinyl chloride in lining concrete sewer pipes for preventing biological corrosion. Int J Env Res 10(2):305–312

Pazoki M, Delarestaghi RM, Rezvanian MR, Ghasemzade R, Dalaei P (2015) Gas production potential in the landfill of Tehran by landfill methane outreach program. Jundishapur J Health Sci 7(4)

Pazoki M, Ghasemzade R, Ziaee P (2017) Simulation of municipal landfill leachate movement in soil by HYDRUS-1D model. Adv Env Technol 3(3):177–184

Pazoki M, Ghasemzadeh R, Yavari M, Abdoli M (2018) Analysis of photocatalyst degradation of erythromycin with titanium dioxide nanoparticle modified by silver. Nashrieh Shimi va Mohandesi Shimi Iran 37(1):63–72

Pazoki M, Pari MA, Dalaei P, Ghasemzadeh R (2015) Environmental impact assessment of a water transfer project. Jundishapur J Health Sci 7(3)

Qasim SR, Chiang W (1994) Sanitary landfill leachate: generation, control and treatment. CRC Press

Shariatmadari N, Abdoli MA, Ghiasinejad H, Alimohammadi P (2010) Assesment of HELP model performance in Arid areas, case study: landfill test cells in Kahrizak. Res J Env Sci 4(4):359–370

Shayesteh AA, Koohshekan O, Khadivpour F, Kian M, Ghasemzadeh R, Pazoki M (2020). Industrial waste management using the rapid impact assessment matrix method for an industrial park. Global J Env Sci Manag 6(2):261–274

Tchobanoglous G, Theisen H, Eliassen R (1993) Engineering principles and management issues. Mac Graw-Hill, New York, 978

Tchobanoglous G, Theisen H, Vigil S (1993) Integrated solid waste management: engineering principles and management Issues. McGraw-Hill

Chapter 5
Leachate Management

5.1 Introduction

Since landfill construction and consequently leachate generation are unavoidable in the waste management process, the main focus is to be on reducing the leachate production as much as possible and also on treating the produced ones in order to eliminate or decline the level of contamination in them up to permissible thresholds before being discharged into the environment.

Leachate management options can be classified into four major groups (Christensen et al. 2005):

1. Leachate evaporation
2. Discharge to the wastewater treatment station
3. Leachate treatment
4. Leachate Recirculating

5.1.1 Leachate Evaporation

Sunlight ends in evaporation of the stored leachate in evaporation marshes with the liner system. Nevertheless, lined leachate evaporation ponds can be covered which depends upon the climatic condition of each location and operational decisions (Di Palma et al. 2002; Shayesteh et al. 2020).

5.1.2 Discharge to the Wastewater Treatment Plant

Some years ago, a common solution was to treat the leachate together with municipal wastewater in the municipal wastewater treatment plant. It was preferred for its

© Springer Nature Switzerland AG 2020
M. Pazoki and R. Ghasemzadeh, *Municipal Landfill Leachate Management*,
Environmental Science and Engineering,
https://doi.org/10.1007/978-3-030-50212-6_5

maintenance and operating costs. In case a landfill is located near a wastewater collection system or available to get linked to the system, leachate can be disseminated to the system and be treated at the wastewater treatment plant. But, pre-treatment of leachate is necessary for reducing organic content before discharging to the sewer (Christensen et al. 2005; Pazoki et al. 2016).

This element has been one of the most debated issues with an increasing concern because of organic inhibitory compounds with low degradability in the course of biological processes and also heavy metals that not only reduce the productivity of the treatment but also increase the concentration of sewage from leaching operations. A potential solution proposes the elimination of nitrogen (caused by leaching) and phosphorus (caused by sewage) from the treatment plants.

In this section, the leachate characteristics, various treatment methods, leachate recirculation, and the designing are discussed.

5.1.3   Leachate Characteristics

The volumetric spill rate and the make-up are the two elements that describe a liquid effluent concerning leachate studies.

The water cycle in a typical landfill is illustrated in Fig. 5.1 (Lema et al. 1988). As depicted, leachate flow rate (E) correlates with precipitation (P), surface run-off (Rin, Rext), and permeation (I) or infiltration of groundwater exuding through the landfill. Conventional methods of structuring landfills (including waterproof covers, applying liners such as clay or plastics) are too primitive to restrict the amount of water entering the pit, and so to reduce the contaminating effects. Climatic conditions can also affect the amount of leachate yield because they have strong influences on the intake of precipitation (P) due to evaporation (EV). Innate characteristics of the waste itself including contents of water and degrees of compaction can affect leachate generation. Considering the fact that compactness can hinder filtration rate, the production rate is higher with less compressed wastes.

Fig. 5.1 Water cycle in a sanitary landfill

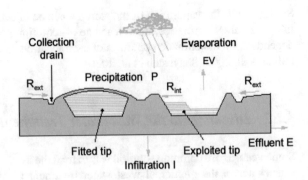

Fig. 5.2 COD balance of the organic ratio across a sanitary landfill

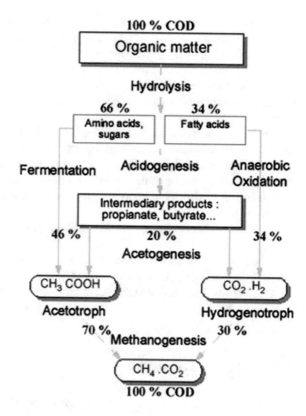

Certain factors can affect the quality of leachates in such landfills: age, precipitation, seasonal variations in the climate, type and composition of the waste (it depends on the living styles of the local residents). Particularly, the age of the landfill has a significant effect on landfill layout that varies accordingly.

Figure 5.2 proposes an anaerobic degradation scheme for the organic material in a sanitary landfill (Lema et al. 1988; Ghasemzade and Pazoki 2017).

In the case of newly-established landfills, biodegradable organic substances exist in large quantities, and when they are accompanied by fast anaerobic fermentation, they produce volatile fatty acids (VFA). The higher level of humidity or water content present within the solid waste can reinforce and catalyze the acid fermentation. This initial period in a landfill lifetime is known as an acidogenic course during which considerable amounts of VFA in the order of 95% of the whole organic content of wastes. As the landfill gets older, the so-called methanogenic phase begins during which microorganisms that produce methane are activated and convert VFA to biogas (CH4, CO2). Then some compounds that are resistant to biological degradation (non-biodegradable) such as humic substances will dominate the organic ratio of the leachate (Lema et al. 1988; Pazoki et al. 2015).

Leachates of landfills are characterized by some fundamental parameters including BOD, COD, BOD/COD ratio, pH, suspended solids (SS), ammonium nitrogen (NH3-N), total Kjeldahl nitrogen (TKN) and heavy metals.

Relevant studies in the available literature have reported widely-varied ranges of the leachate taken from different sanitary landfills. The time of the landfill and therefore the amount of solid waste stabilization has an important impact on water/leachate features.

Albeit leachate compounds may changes considerably within the continuous aerobic, acetogenic, methanogenic, stabilization phases of the waste evolution, three types of leachates have been defined according to landfill age. The existing relationship between the age of the landfill and the organic material composition may provide useful yardsticks to choose a proper treatment process (Lema et al. 1988; Qasim and Chiang 1994).

The goal of this section is to present an all-out review of landfill leachate treatment processes.

Besides, their treatment practices are evaluated based on COD, NH3-N and heavy metal, elected data on pH, the dose required, and strength of wastewater with a view to COD, NH3-N, and heavy metal concentration and also treatment efficiency (Qasim and Chiang 1994).

5.2   Leachate Treatment

Leachate treatment by means of biological or physical/chemical processes and choices is selected regarding the concentration of polluting material in leachate that needs to be omitted.

There are a few criteria that are of great significance for the kind of treatment technology that has to be applied, mainly COD and AOX, furthermore Nitrogen and BOD$_5$. The first two criteria require respectively a more all-inclusive treatment technology and a mixture of different treatment methods.

New methods intended for leachate treatment can be classified as follows:

1. Chemical and physical methods
2. Biological methods
3. Filtration methods. But because of the existence of high COD, hybrid pilots are used for the treatment of landfill leachate.

5.2.1   Chemical and Physical Methods

Chemical/physical processes are carried out either as supplementary operations integrated with the main treatment line (as a pretreatment or final purification) or in order to treat a specific contaminant (stripping for ammonia).

5.2.1.1 Flotation

Up to this point, a few types of research have been carried out to examine the application of flotation in the treatment of landfill leachate. In recent times the use of flotation in the column, as a post-treatment phase for omitting residual humic acids (non-biodegradable compounds) from simulated landfill leachates has been reconsidered. Under optimized conditions, almost 60% of humic acid removal has been achieved (Qasim and Chiang 1994; Wiszniowski et al. 2006).

5.2.1.2 Activated Carbon Adsorption

In most cases, the activated carbon adsorption method has illustrated eminence in the removal of a vital quantity of organic amalgams and ammonium nitrogen from the leachate samples. Trying to tackle temporal fluctuations in varying strength and composition of landfill leachate on one hand, and ameliorating the single-phase adsorption procedure, on the other hand, development of a collaborated multistage treatment that combines adsorption procedure with numerous complementary approaches has lately given immense attention. Hence, some researchers have made new hybrid procedures that use activated carbon with other processes such as the combination of ozone-GAC adsorption treatment, PAC-GAC and biomimetic fat cell (BFC) coagulation/flocculation and powder activated carbon adsorption nanofiltration and microfiltration–PAC hybrid process (Renou et al. 2008; Delarestaghi et al. 2018).

Each type of these procedures has a specific pilot. For example, in the integrated ozone-GAC adsorption treatment, initially, an ozone contactor is used in lieu of pre-treatment and then a retention tank and at last GAC adsorption is used.

The application of activated carbon with ozone appears to be very appealing today as a result of the advantages of using this combination for leachate treatment. The simplicity of the system and its ability to adopt different power and composition of landfill leachate vary with seasonal changing.

The usage of activated carbon adsorption (GAC or PAC) is effective for the removal of non-biodegradable compounds from leachate, but not for NH3–N. More than 90% of COD was removed with its concentration ranging from 940 to 7000 mg/L. Nevertheless, the need for a repetitive regeneration of activated carbon column and the high cost of GAC may limit its application for the treatment of landfill leachate in developing countries (Foul et al. 2009).

But if GAC and ozone are used with combination procedures as it has done by NH3-N, it could be omitted.

As can be viewed, the mixture of ozone- GAC adsorption using ozone-modified GAC had the greatest removal enforcement for COD (86%) and/or NH3-N (92%) when compared to ozonation alone (COD: 35%; NH3-N: 50%) at the same preliminary COD and/or NH3-N consolidations of 8000 and 2620 mg/L.

5.2.1.3   Coagulation–Flocculation

Several studies have been reported on the investigation and surveying of coagulation-flocculation for the treatment of landfill leachates, aiming at process optimization, i.e., selection of the most appropriate coagulant, identification of optimal experimental situations and assessment of pH effect. Recent works clearly reveal that iron salts are more merited than aluminum ones, resulting in sufficient chemical oxygen demand (COD) cuts (up to 50%), whereas the corresponding values in case of aluminum or lime addition were intermediate (between 10 and 40%). Nevertheless, the mixture of coagulants or the addition of flocculants together with coagulants may enhance the floc-settling rate and so the process enforcement (COD abatement up to 50%) (Qasim and Chiang 1994; Amokrane et al. 1997).

5.2.1.4   Chemical Precipitation

During the operations arranged for leachate treatment, a pre-treatment phase in the form of chemical discarding is necessary in order to attenuate the power of ammonium nitrogen ($NH^+_4 - N$). For the purposes of an experiment conducted by Li et al., they precipitated ammonium ions as magnesium ammonium phosphate (MAP) with the annexation of $MgCl_2 \cdot 6H_2O$ and $Na_2HPO_4 \cdot 12H_2O$ with an $Mg/NH_4/PO_4$ ratio of 1/1/1 at a pH of 8.5–9. Applying this procedure for 15 min confined the concentration of ammonium from 5600 to a range of 110 mg L^{-1} (Qasim and Chiang 1994; Zhang et al. 2009).

5.2.1.5   Air Stripping

Air stripping involves transferring of volatile components of a liquid into an air stream and is used for the purification of groundwaters and wastewaters containing volatile compounds. Since landfill leachates contain large quantities of ammonium nitrogen, air stripping can be effective in removing this contaminant and minimizing wastewater toxicity (Campos et al. 2013).

5.2.1.6   Electrocoagulation

The result of COD removal efficiency by Fe-electrode is different than Al-electrode. One reason for this phenomenon is the number of electrodes used. The main problem of EC is economical issues. This process is perfect for some situations including lack of space. As the majority of landfills are situated out of the cities, they have also enough nearby free land, but normally EC is not feasible for the treatment of leachate for landfills out of the cities (Campos et al. 2013; Pazoki et al. 2018).

Therefore, some new methods are investigated to improve the efficiency of this process; for example, electro-Fenton or hybrid process which combines EC with another kind of process like nanofiltration.

Electro- Fenton will be presented here and nanofiltration will be introduced in the coming paragraphs.

Electro –Fenton is a new process that is used in the new research.

In general, Fe^{2+} will be rapidly used up and the production of hydroxyl radical has to rely primarily on a H_2O_2/Fe^{3+} reaction or Fenton-liked process which is much slower; thus, the pollutant oxidation speed slows down.

As far as EC is relatively a new method, so a comprehensive treatment procedure with EC couldn't be found in the research literature. Nevertheless, there are other reasons too; for instance, landfills are typically built on suburbs where there is an absence of infrastructures, so other kinds of processes that are more useful and cheaper can be used.

It appears that EC is a new way of treatment, so a full-scale treatment of this procedure with EC could not be found. It has some logic for example landfills are built in the suburbs of the city so they can use other kinds of procedures that are more useful and cheaper.

But in ordinary conditions, it is suggested to use a type of pre-treatment method like coagulation.

Due to the fact that, if the wastewater with a great amount of COD is needed for this procedure, the electrode will be ended sooner and this will make exorbitant costs. So it is better to utilize EC with other procedures to make it economically merited.

5.2.1.7 Chemical Oxidation

Oxidation can be accomplished by various methods that are most common in applying strong oxidants (e.g. O_3 and H_2O_2), irradiation (e.g. ultraviolet (UV), ultrasound (US) or electron beam (EB)), and catalysts (transition metal ions or photocatalyst).

Even though many of the previous researchers are utilizing ozonation because it demonstrated the effectiveness in eliminating COD (reduction is about 50–70% in most cases), most of them only utilized this procedure as a tertian treatment before discharge in the environment. Sometimes the treatment efficiency on stabilizing leachate has been moderated. Among all procedures, Fenton's process seems to be the best trade-off because the process is technologically plain, there is no mass transfer obstruction (homogeneous nature) and both iron and hydrogen peroxide is cheap and non-toxic. But Fenton's process requires low pH and modification of this criterion is essential (Loizidou et al. 1993; Ghasemzadeh et al. 2017).

5.2.2   Biological Treatment

Biological treatment methods are processes whereby microbes are used to eliminate or at least reduce the toxicity of a waste stream. Normally, the biological treatment of predominantly aqueous wastes such as leachate is accomplished in specially designed bioreactors. A suitable culture of the micro-organisms or microbial association, either aerobic or anaerobic, is chosen. Biological treatment has gained wide support because of providing a reliable, convenient, and cost-effective procedure for purification of leachates with large contents of BOD. Biodegradation is accomplished by microorganisms that may lead to the generation of two ranges of products depending on the present conditions: they may convert organic compounds into carbon dioxide and sludge under aerobic conditions, or to biogas (mostly contains CO_2 and CH_4) under anaerobic conditions. Biological processes have been confirmed as efficient procedures capable of eliminating nitrogen-contained substances from young leachates with a high BOD/COD ratio in the order of > 0.5 (McArdle et al. 1988; Chian and Dewalle 1976).

5.2.2.1   Aerobic Biological Treatment Processes

Most leachate treatment plants that have been established in various regions around the world involve a phase for aerobic biological treatment. The superiority of this procedure over other methods resides in its unique capability of degrading pollutants rather than transforming them. Actually, the basic principles of the aerobic biological process can be traced to suspended-growth biomass including ventilated lagoons, conventional activated sludge processing, and sequenced-batch reactors (SBR) that have been contemplated of, modified, and improved. Recently, attached-growth systems have gained in popularity that involve moving-bed biofilm reactor (MBBR) and biofilters. Membrane bioreactor that has integrated membrane separation technology with aerobic bioreactors, has introduced a novel approach to leachate treatment lately (McArdle et al. 1988).

Most aerobic biological treatment systems with various arrangements share common characteristics that will be discussed in detail in the following sections.

(a)   Treatment of COD and BOD;
(b)   Treatment of ammoniacal-N;
(c)   Treatment of trace organic and other compounds.

(a)   **Treatment of COD and BOD**

Considering the requirements of aerobic biological treatments, the 5-day biochemical oxygen demand (BOD5) test is not able to provide enough amount to enable degradation of organic compounds in the treatment plants that contain a population of bacteria that have been adapted to this condition. A standard bacterial seed is used in the BOD5 test with a finite incubation period of 5 days. The COD value is usually utilized for meeting the purposes of design in spite of the fact that the 20-day test (BOD20) might appear to be more helpful. The outcome of well-designed treatment plants indicates elimination efficiency for getting rid of organic mixtures that is much higher than the rate estimated by the 20-day BOD. The obtained knowledge form wide-range experiments in the context have elucidated that contents of remaining, non-degradable COD in emissions of treatment processes do not necessarily correlate with the concentration of organic elements in crude leachate, whereas it may affect the concentration of ammoniacal-N by contrast. Perhaps, we may attribute that to its emancipation during degradation of wastes or to its generation as a result of the nitrification process.

(b) Treatment of ammoniacal-N

Nitrification

Nitrification is the biological oxidation of ammonia or ammonium to nitrite followed by the oxidation of the nitrite to nitrate. Nitrification is an aerobic process performed by small groups of autotrophic bacteria that acquire their energy from the oxidation reaction and consumes inorganic carbon as an essential food source. The nitrification process is a type of oxidation that is accomplished in two phases, and a distinct group of bacteria is responsible for each phase. Bacteria belonging to the genus Nitrosomonas carry out the first phase i.e. oxidation of ammoniacal nitrogen to nitrate nitrogen. Species of Nitrobacter are responsible for the second phase through which the nitrate-nitrogen is further oxidized and changed to nitrate nitrogen.

Denitrification

Biological denitrification refers to the process of reducing nitrate-nitrogen to nitrogen gas by facultative heterotrophic organisms that acquire the needed energy from organic carbon. These organisms are capable of operating under both aerobic and anoxic conditions using molecular oxygen and nitrate nitrogen as the electron acceptor respectively. The anoxic phase refers to a region which only contains chemically-combined oxygen, and bacteria can also consume it in the form of nitrate-nitrogen or nitrate nitrogen. It is noteworthy that such a situation must clearly be differentiated from anaerobic one that refers to a circumstance of complete lack of accessible oxygen.

(c) Treatment of trace organic and other compounds

5.2.2.2 Suspended-Growth Biomass Processes

Lagoon

Traditional approaches that were initially developed during the 1970 and 1980s, focused on creating ventilated lagoons. Lagoons of that time were typically large and 1–2 m deep with a special design making their appearances similar to natural wetlands surrounded by vegetation. Aerated lagoons have typically been viewed as an effective and low-cost method for removing pathogens, organic and inorganic matters. Their low operation and maintenance costs have made them a common choice for wastewater treatment, particularly in developing countries since there is a little need for specialized knacks to run the system. Overall rates of eliminating N, P, and Fe provided by this system ranged over 70% in the case of thinned leachate. In spite of the fact that lagoons can provide sufficient and cost-effective treatments for leachate, more demanding conditions may not be adequately dealt with by making lagoons. One of the plainest forms of on-site treatment of landfill leachate is aerated lagoons, with aeration by means of surface aerators or by diffuse bubble aeration. Treatment occurs via chemical and biological oxidation. However, if the aerated lagoon way of treatment is to be adopted as a full-scale means of leachate treatment then the consequences of changing the Hydraulic Retention Times (HRT) must be fully assessed. In the most practical full-scale applications, the HRT will be determined by the flow rates of leachate into the reactors with shorter HRT in winter months and longer reservations in summer. Aerated lagoons, with properly long HRT, omitted large extents of COD and ammonia.

Activated Sludge Processes

The most widespread biological processing is activated sludge, which is a suspended-growth procedure that utilizes aerobic microorganisms to biodegrade organic contaminants in leachate. With standardized activated sludge treatment, the leachate is aerated in an open tank with diffusers or mechanical aerators. Activated sludge is made up of suspended biological flocs that are matrices of microorganisms, nonliving organic matter, and inorganic materials. The activated sludge or biological flocs mix with the waste stream; oxidize the organic substances in the wastewater while oxygen exists for bio-oxidation and nitrification reactions or in the absence of oxygen for denitrification reaction. These processes are widely-used for treating household sewage or co-treatment of leachate and sewage. Whereas, recent studies conducted in this filed have confirmed the insufficiency of this approach in dealing with treating landfill leachates. The fundamental basis behind all activated sludge processes is that as microorganisms grow, they form particles that clump together. These motes, which are referred to as floc, are permitted to settle to the undermost of the tank, which ends in a rather clear liquid free organic material and suspended solids (Qasim and Chiang 1994).

The screened wastewater is comingled with changing quantities of recycled liquid that involves high proportions of organisms that are extracted from a secondary tank and turns into a product that is named mixed liquor. The next step for the amalgam is to stir and inject it with large amounts of air to provide oxygen and hold the solids in suspension. After a period of time, the mixed liquor flows to a clarifier where it is permitted to settle. During this settling, a portion of the bacteria is removed and the partially cleaned water runs on for additional treatment. The final settled solids, the activated sludge, are then returned to the first tank to begin the process again. Addition of PAC to triggered sludge reactors augmented nitrification efficiency in the biological treatment of landfill leachate (Qasim and Chiang 1994; Renou et al. 2008).

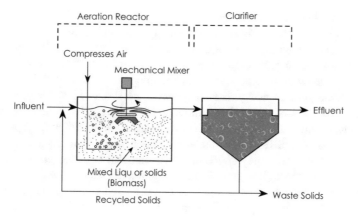

Fig. 5.3 Typical activated sludge process

Despite its undeniable efficiency in removing organic carbon, nutrients, and ammonia contents, some drawbacks have been characterized:

– Inadequate sedimentation of sludge that requires longer ventilation,
– The energy-intensive process leading to excess sludge,
– Microbial inhibition thanks to the high amount of ammonium-nitrogen power (Fig. 5.3)

Sequencing Batch Reactor (SBR)

As this system submits an operation regime which is in alignment with concurrent organic carbon oxidation and nitrification, it is ideally compatible with nitrification–denitrification procedures. The Sequencing Batch Reactor (SBR) treatment procedures have been devised as a readily-automated, extended aeration system that is specifically well suited to the greater organic strength and consolidations of ammoniacal-N in landfill leachates. The greater amount of the main SBR tank makes for efficient aeration, high rates of dilution of incoming leachates and high resistance to shock loading. The most significant property of SBR, as a critical concern in the aerobic treatment of household leachates, is its wide-range flexibility in the process, since it highly varies in quantity and quality. Specially-designed SBR systems that have been tailored for certain loading rates, for either ammoniacal-N or organic pollutants, require substantial flexibility to effectively handle either strong leachates in small quantities or weak leachates in large quantities. Since characteristics of leachates may vary over time, such a feature plays a significant role to ensure sustaining the most effective performance. Sequencing batch reactor processes actually represent a very elementary form of treatment process systems known as fill and draw, similar to the household washing machine operation. The entire treatment is accomplished in one reactor. Various sequencing batch reactor procedures have been developed for a wide range of environmental applications

including portable/industrial water treatment, municipal/industrial wastewater treatment, and solid waste handling and treatment. The sequencing batch reactor processes
can be biological, physicochemical or biological physicochemical. Biological treatment of landfill leachate usually results in low nutrient eliminations due to the high
chemical oxygen requirement (COD), high ammonium-N content and the presence
of toxic blends such as heavy metals. However, no studies have been reported on
nutrient (COD, NH4-N, and PO4-P) elimination from landfill leachate by means of an
SBR. Thus, Uygur et al. have the first report on nutrient elimination from pre-treated
landfill leachate in an SBR. Three, four and five-step SBR operations were contrasted
in the first phase of the study concerning their nutrient omission efficiencies from
the pre-treated leachate (Renou et al. 2008; Timur and Özturk 1999).

A traditional SBR process is similar to an activated sludge process. The main
advantages of a sequencing batch reactor procedures are as follows (Qasim and
Chiang 1994; Renou et al. 2008):

1. Developed effluent qualities,
2. The omission of separate clarifiers and sludge return pumps,
3. Increased settling region,
4. A completely quiescent settling environment,
5. Demand controlled energy consumption,
6. Short-circuiting removed,
7. A special ability to handle extremely high organic and hydraulic shock burdens,
   and
8. The capacity to equalize flows and load.

Attached-Growth Biomass Systems

The attached-growth or fixed film system in the context of aerobic biological treatment involves growing bacteria attached to the surface of an inactive medium, typically plastics, or to the circulating rotor of an RBC system. Moreover, certain alternative approaches have been also proposed with both advantages and disadvantages
for each method. State of art systems that are commonly known as attached-growth
biomass engaging biofilms has been developed to compensate for major challenges
of sludge bulking and insufficient detachability in conventional aerobic systems. The
superiority of these systems over other systems resides in allowing applicability even
in the absence of active biomass. Furthermore, in comparison with suspended-growth
systems, they are less affected by low temperatures and inhibition thanks to higher
amounts of nitrogen.

Trickling Filters

Trickling filter consists of a fixed biological bed of rock or plastic media on which
wastewater is used for aerobic biological treatment. Biological slimes shape the
media which assimilates and oxidize materials in the wastewater. This procedure
has been investigated for the biological nitrogen lowering from municipal landfill
leachate. Biofilters remain an interesting and charming option for nitrification due
to low-cost filter media. Trickling filters enable organic matters in the wastewater to

be absorbed by microorganisms (aerobic, anaerobic, and facultative bacteria; fungi, protozoa, and algae) attached to the medium as a biological film or slime layer (0.1 to 0.2 mm). As the wastewater flows over the filters, microorganisms of water gradually connect themselves to the rock, slag, or plastic surface and form a film. As the layer thickens through microbial growth, oxygen cannot infiltrate the medium face, and anaerobic organisms develop. As the biological film continues to rise, the microorganisms near the surface lose their ability to cling to the medium, and a portion of the slime layer falls off the filter. R. Matthews et al. also examined treating landfill leachate using passive aeration trickling filters. The trickling filter system reported in their study proved the capability of treating a variety of landfill leachates typical of operational and closed sites with broadly consistent performance. The treatment rates reported are lower than those for some other treatment systems but they were attained without their additional operational sophistication, monitoring requirements and costs. For some strong leachates, $NH_4 + - N$ and nitrogen oxidation procedures were temporally decoupled. The extent to which this phenomenon is attributable to transient N storage within microbial films, to the presence of undetected N-forms within massive leachate, or a combination of these elements, needs further examination (Fig. 5.4) (Qasim and Chiang 1994; Renou et al. 2008).

Human-made wetlands

Creating artificial marshes and reed beds is an effort to imitate natural processes through which polluted water is treated when passes across wetlands. Wastewater treatment in such systems is accomplished by the degradation of organic elements (BOD and COD), oxidation of ammoniacal-N, elimination of suspended solid particles, and also reducing compressions of nitrate and phosphorus to a lesser extent. Arrangements within the treatment process are sophisticated and usually involve bacterial oxidation, filtration, nitrification, and chemical precipitation. This procedure was introduced in 1985 following a series of inspections carried out by Water Industry employers of several systems that have been established in Germany.

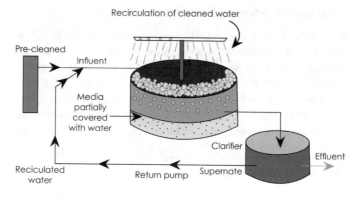

Fig. 5.4 Trickling filter

Initially, a fundamental model of wetlands was put into operation, then by introducing successive models, some modifications have been brought about in their performances (Renou et al. 2008; Mulamoottil et al. 2018). These systems can be broadly categorized into two types:

1. Horizontal flow;
2. Vertical flow.

Most flourishing applications in both types engage subsurface flow through a medium composed of gravel and sand in which the reeds are grown. By doing this, the surface flow will be avoided disregarding the main treatment surfaces.

Horizontal flow reed beds/human-made wetlands

Leachate treatment is carried out with the aid of placed gravels of 10 mm in size to a depth of 600 mm. An inclined plane that is built across the basal level allows draining from the bed, provided that maximum gravel depth does not exceed 1 m. Horizontal-flow beds are particularly useful for direct treatment of relatively thin leachates, for example from old or closed landfills. In such conditions, the operation will be more convenient because of low masses of ammonia and the distantness of the sites, so this system will be the most favorable alternative. However, in the case of stronger leachates, the horizontal flow reed bed can only appear as an ancillary phase in addition to the main biological process, immediately before the final stage of releasing contaminants to canals.

Treating thin leachates directly/polluted groundwaters

Horizontal-flow reed beds have been demonstrated to be effective in treating thin leachates (i.e. ammoniacal-N < 25 mg/l). Their efficacy in minimizing harmful contents of residual BOD suspended solids, iron, and small concentrations of organic mixtures like mecoprop (methyl chlorophenoxy propionic acid (MCPP), a common herbicide) has been experimentally approved as an appropriate method dealing with old and landfills (Pazoki et al. 2017; Pazoki et al. 2015) (Fig. 5.5).

Refining biologically pre-treated leachates

Refining the leachates that have previously treated by biological processes has become a common procedure in most landfills in order to improve the emitted wastewater being released to surface with the aid of horizontal flow reed beds. A model developed by Hampshire County Council for treating leachate from Efford landfill is an example of such procedure. Treating capacity of an SBR system is approximately 150 m^3 of thick leachate per day. The resultant emissions pass through a reed bed before being released into a small rural STW.

Moving-bed biofilm reactor (mbbr)- suspended-carrier biofilm reactor (scbr)- fluidized bed reactor)

The fundamental structure of MBBR consists of suspended porous carriers that are continuously kept moving inside a ventilating tank, and they absorb the active biomass being raised as a biofilm that is attached to their surfaces. This procedure

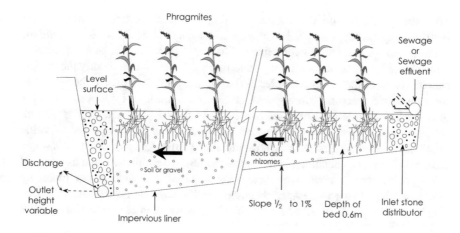

Fig. 5.5 Typical arrangement for a horizontal-flow reed bed treatment system

has some advantages over conventional methods: higher concentration of biomass, shorter time needed for sludge settlement, lower sensitivity to poisonous compounds, and integrative elimination of organic contents and ammonia by a single process (Fig. 5.6).

Rotating biological contactors (RBC)

The rotating biological contactor is an instance of a biological filter (attached growth) technology. It consists of circular plastic discs mounted centrally on a usual horizontal shaft. These discs are almost 40% submerged in a tank containing wastewater and are slowly rotated by either a mechanical or a compressed air drive. Microorganisms from the wastewater adhere to the plastic disc surfaces and, within 1–4 weeks from start-up, form a biofilm alternating from 1 to 2 mm in thickness. This biological growth assimilates organics from the wastewater passing over the surface of the disc and is responsible for most of the treatment which occurs. When the disc rotates out of the wastewater, the biofilm becomes exposed to air and is oxygenated, thereby

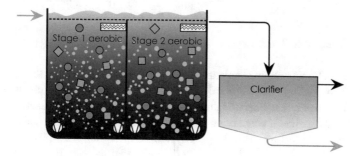

Fig. 5.6 Moving-bed biofilm reactor

maintaining aerobic situations. After reaching a critical thickness, portions of the biofilm slough off the discs. RBCs allow a wider range of flexibility for treating the leachates compared with trickling filters. Moreover, by making certain adjustments on the main arrangements of the tank and rotor, it is possible to modify the mixing specifications of the system in order to confine plug flow and enable the system to run as a mixed reactor. Such a system allows fast dilution of infiltrating leachates generated by ventilation processes like SBR. In the RBC system treating landfill leachate, the ammonium load was a significant parameter affecting the nitrification efficiency and its main products. At an ammonium load of $1.92 \, g \, NNH_4/m^2 \, d$, a single RBC was enough in obtaining complete nitrification; nevertheless, at a higher load $- 3.6 \, g \, N\text{-}NH_4/m^2 \, d$, a two-stage system was required. The increase of ammonium load more than $4.79 \, g \, N\text{-}NH4/m^2 \, d$ caused a fall in nitrification efficiency to 70%. The bacteria populations altered with changing ammonium concentrations (Qasim and Chiang 1994; Renou et al. 2008).

Advantages

Advantages of RBCs include:

- Short contact periods are required because of the large active surface.
- RBCs are suitable for handling a wide range of flows.
- In general, sloughed biomass has good settling characteristics and can readily be segregated from the waste stream.
- Operating costs are negligible because little skill is required in plant operation.
- Short retention time.
- Low power needs.
- The omission of the channeling to which usual percolators are susceptible.
- Low sludge production and excellent procedure control.

Disadvantages

Disadvantages of RBCs include:

- A need for covering RBC units in northern climatic conditions to preserve against freezing.
- Shaft bearings and mechanical drive units need recurring maintenance (Fig. 5.7).

5.2.2.3 Anaerobic Treatment

Anaerobic treatments allow terminating the process which begins in the tip making it effective in dealing with higher concentrations of organic-based emissions that are typical for newly-established landfills. Contrary to aerobic processes, anaerobic treatments are capable of conserving energy and producing small amounts of solids; however, reaction rates are so low in these systems. Furthermore, it is possible to make use of the generated CH_4 for heating the degrader that its operational temperature range is at 35 °C, so desirable conditions to meet the predefined goals can be easily

Fig. 5.7 Rotating Biological
Contactor (RBC)

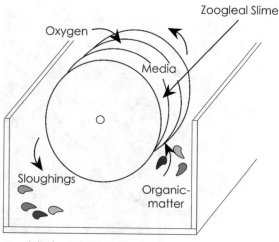

Wastewater holding tank

provided. Anaerobic degradation suffers from some problems that can be summarized as follows:

This method is not only unable to eliminate ammoniacal-N but also can augment concentrations of this harmful pollutant. It is required to be provided with an auxiliary aerobic process.

In order to construct a treatment process in aerobic conditions to be self-sufficient in terms of energy, the content of COD in crude leachate should not go below 10,000 mg/l. The acetogenic stage that is common in most modern landfills usually has short life spans.

The highest efficacy of an aerobic process is only achieved when it is embedded inside the landfill body itself to be provided with desirable and stable temperatures. A well-monitored system for acetogenic leachates recirculation can meet the requirements, then final landfill gas will be collected by the existing systems.

Suspended-growth biomass processes

Up-flow anaerobic sludge blanket (UASB) reactor

UASB is a totally novel procedure in the field of anaerobic treatment that is capable of providing high treatment applicability with short hydraulic retention time. The fluid waste that is to be treated enters the reactor at the base then moves up through a sludge blanket. The blanket is composed of granular anaerobic bacteria that have a high sedimentation rate, so they can be retained in the reactor. As the waste flows up and moves through the sludge blanket, the output fluid is discharged from the top of the reactor. Blending and mixing operations are facilitated by the generation of anaerobic gases and the upward flow of the fluid. Similar to other anaerobic treatments, UASB is also too inefficient to be used for dealing with leachates. UASB reactors show higher efficiency than their equivalent systems regarding the capacity

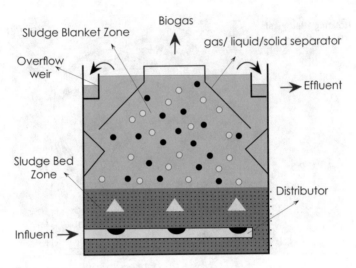

Fig. 5.8 Up-flow anaerobic sludge blanket (UASB) reactor

of handling large volumetric organic rate value. The operational temperature in the case of UASB reactors ranges from 20 to 35 °C (Fig. 5.8).

UASB is very efficient in treating any organic loading wastewater. Organics are degraded and produce biogas as a source of energy renewal. A two-stage up-flow anaerobic sludge blanket and sequencing batch reactor system was used to treat land-fill leachate for the advanced omission of nitrogen and COD. In order to access the total nitrogen (TN) removal efficiency and to reduce the COD requirement for deni-trification, the raw leachate with recycled SBR nitrification supernatant was pumped into the first-stage UASB to achieve simultaneous denitrification and methanogen-esis. The results showed that UASB played an important rule in reduce COD and SBR further enhanced nutrient removal efficiency.

Attached-growth biomass processes

Anaerobic filter

The anaerobic filter is a high rate mechanism that gathers the advantages of other anaerobic systems and minimizes the disadvantages. In an up-flow anaerobic filter, biomass is preserved as biofilms on a support material, such as plastic rings (Qasim and Chiang 1994; Henry et al. 1987).

Hybrid bed filter

It is made up of an up-flow sludge blanket at the bottom and an anaerobic filter on top. This device acts as a gas–solid separator and augments solid's retention without causing channeling or short-circuiting. Enhanced performances of such a process originating from the maximization of the biomass concentration in the reactor.

Fig. 5.9 Fluidized bed reactor

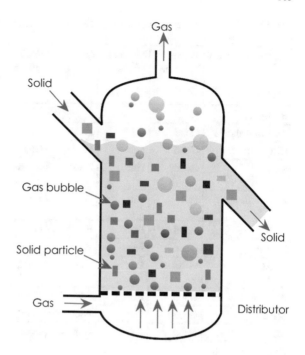

Fluidized bed reactor

A fluidized bed reactor (FBR) is a kind of reactor that can be used to carry out a series of chemical reactions. In this reactor, a fluid (gas or liquid) is passed through a granular solid material at high enough velocities to suspend the solid and cause it to act as though it were a fluid. This process, known as fluidization, have important advantages to the FBR. As a result, the fluidized bed reactor is now used in many industrial applications. As reported by several experiments carried out in the field of carbon-assisted fluidized beds, the integration of biodegradation and absorption into a single process allows the omission of various organic compounds. Imai et al. rated the biological activated carbon fluidized beds higher than traditional approaches for their efficiencies in treating leachates from old landfills (Qasim and Chiang 1994; Gulsen and Turan 2004) (Fig. 5.9).

5.2.3 Summary

The leachate treatment procedures were compared in terms of treatment efficiency, space usage, installation, and operational cost. Some details of the full-scale treatment plants such as operation year, the capacity of the plant, country of location, substances removed and treatment procedures applied were also searched for and listed. Table 5.1 is a comparison of the techniques in terms of treatment efficiency, space usage, installation, and operational cost (Madu 2008).

Table 5.1 Comparison base on treatment efficiency, space utilization, installation, and operational costs

Treatment process	Young Leachate	Medium leachate	Old leachate	Space utilization	Installation and operational cost	Requiring less skilled personnel
Biological						
Activated sludge	Good	Fair	Poor	Poor	Expensive	No
RBC	Good	Fair	Poor	Good	Expensive	Yes
SBR	Good	Fair	Poor	Good	Less expensive	No
Reed Beds	Fair	Fair	Good	Poor	Less expensive	Yes
BAF	Good	Fair	Fair	Good	Expensive	Yes
Lagoons	Good	Fair	Poor	Poor	Expensive	Yes
UASB	Good	Fair	Fair	Good	Less expensive	Yes
AF	Good	Fair	Fair	Good	Expensive	Yes
MBBR	Good	Fair	Poor	Poor	Expensive	No
MBR	Good	Fair	Fair	Poor	Expensive	No
Physicochemical						
Coag. & Flocculation	Poor	Fair	Fair	Fair	Less expensive	No
Precipitation	Poor	Fair	Poor	Fair	Less expensive	No
Adsorption	Poor	Fair	Good	Good	Less expensive	No
Flotation	Poor	Fair	Fair	Poor	Expensive	Yes
Chem. Oxidation	Poor	Fair	Fair	Good	Expensive	No
Ammonia stripping	Poor	Fair	Fair	Poor	Expensive	No
Membrane process						
Microfiltration	Poor	Poor	Poor	Good	Expensive	Yes
Ultrafiltration	Fair	Fair	Fair	Good	Expensive	Yes
Nanofiltration	Good	Good	Good	Good	Expensive	Yes
Reverse Osmosis	Good	Good	Good	Good	Expensive	Yes

5.2.3.1   New Treatments

Newly-developed methods in landfill treatments are mostly associated with membrane processes including Micro Filtration, ultrafiltration, nano-filtration, and reverse osmosis.

5.2.4 Micro-Filtration (MF)

Whenever there is an urgent need for the omission of colloids and suspended solids, for example, any pre-treatment prior to other membrane processes or as a collaborative procedure along with chemical treatment, perhaps MF is the only option that comes first. Nevertheless, it's not possible to use it as an independent method that can stand alone.

5.2.5 Ultra-Filtration (UF)

UF is effective to eliminate the macromolecules and the particles, but it is strongly contingent on the type of material making the membrane. UF may be used as an instrument to fractionate organic matter and so to evaluate the preponderant molecular mass of organic pollutants in given leachate. Tests with membrane permeate may give data about recalcitrance and toxicity of the permeated fractions as well. Except, UF was eliminated as a primary tool for treating landfill leachate due to drastic existing regulations. These authors used membranes close to nanofiltration, leachate had low organic matter content and local water standards were not so strict. However, suggested that UF might prove to be effective as a pre-treatment process for reverse osmosis (RO). UF can also be used to remove the larger molecular weight components of leachate that tend to foul reverse osmosis membranes UF has recently used as a biological post-treatment procedure dealing with landfill leachates. Several integrated arrangements of multiple processes like activated sludge–ultrafiltration–chemical oxidation and activated sludge–ultrafiltration–reverse osmosis that is used complementarily have been evaluated so far. The findings indicate that UF by itself is capable of segregating the organic substance by 50%.

Additionally, UF membranes have applied in full-scale membrane bioreactor plants, and have resulted in desirable outcomes. It also yields considerable levels of leachate treatments in landfills.

5.2.6 Membrane Bioreactors

Recently, the wastewater treatment industry has adopted an integrative approach that combines membrane separation technology with bioreactors. Such an arrangement creates very compact systems that operate with high biomass concentration, low sludge production rates, and superior quality of emissions.

Membrane bioreactors have been successfully used for several industrial wastewater treating applications and been also tailored to meet leachate treatment requirements by some plants.

Nevertheless, only a few studies have focused on applying membrane bioreactors to landfill leachate refinement. Contrary to conventional procedures, organisms such as nitrifiers or organisms of biodegradable materials are not removed from the system, so the process has no activity loss.

5.2.7   Nano-Filtration (NF)

NF is a multi-purpose technology that is able to accomplish various objectives associated with water quality such as enabling control measures over organic, inorganic, and microbial pollutants.

A high rate of rejecting sulfate ions and dissolved organic elements on one side and the low rate of rejecting chloride and sodium on the other side can minimize the concentration.

A few studies have dealt with applying NF to leachates treatment applications. NF is able to eliminate 60–70% COD and 50% ammonia regardless of membrane material and geometry (flat, tubular, or spiral wounded), supporting average velocity of 3 m/s and transmembrane pressure within the range of 6 – 30 bar. Physical techniques have successfully been used in integration with nano-filtration with desirable outcomes in terms of eliminating COD from the leachate (70–80%). However, the optimum application of membrane technology requires well-calibrated monitoring measures over membrane spoiling. A wide range of components may lead to membrane spoiling: dissolved organic and inorganic substances, colloidal and suspended solids.

5.2.8   Reverse Osmosis (RO)

RO appears to be the most promising and effective approach to leachate treatments among its other counterparts. A wide range of studies has assessed the efficiency of RO for the eradication of contaminants from landfill leachates in both synthetic environments of labs and realistic industrial sites. The rejection coefficient regarding the COD factor and contents of heavy metals has been reckoned at over 98 and 99%.

Considering strict regulations enforced on landfill management in recent times, previously-used systems such as aerobic or anaerobic biological methods seem to become inapplicable for meeting the developed requirements associated with wastewater treatments. Thus, membrane procedures, and particularly RO, have gained much more supports as the most efficient solutions that are capable of:

- Realizing perfect purification (rejection rates of 98–99%)
- Obviating the challenge of water pollution.

As illustrated by Fig. 5.10, the French case can represent a global trend toward replacing biological treatment plants with pressure-driven membrane processes.

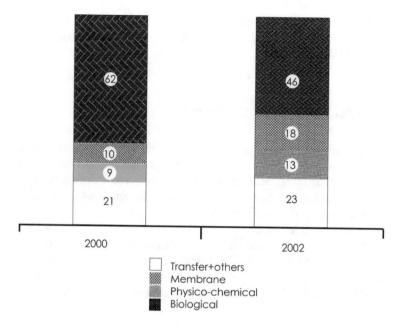

Fig. 5.10 Landfill leachate treatment distribution, in France

However, the feasibility of RO closely depends on monitoring treatment costs and choosing appropriate pre-treatment procedures focusing on minimizing membrane spoiling. A certain fraction of wastes inevitable has to be released as process residue that still has remained a critical challenge in relation to landfills. Carrying the residues to combusting plants to be burnt may offer a possible solution, but it also necessitates additional costs (Renou et al. 2018).

As can be viewed in the previous section, even though leachate treatment is a sophisticated expensive process, another leachate management calculating factor with lower expenses and more privileges is leachate recirculation which will be talked about in the next sections.

5.3 Leachate Recirculating

Leachate recirculating is one of the multiple techniques used to manage leachate from landfills. The main goal of leachate control is to prevent unhindered spreading. Leachate should always be collected, treated or contained before it is released into the environment. During leachate recirculation, leachate is returned to a lined landfill for re-infiltration into the MSW. This is considered as a method of leachate control because as it continues to flow through the landfill, leachate is treated by means of biological processes, precipitation, and sorption. This mechanism also benefits the landfill by increasing the moisture content which in turn augments the rate of biological degradation in the landfill, the biological stability of the landfill, and the rate of methane recovery from the landfill (Renou et al. 2008; Pazoki et al. 2014; Pazoki et al. 2012).

5.3.1 Definition of Leachate Recirculation

Leachate recirculating involves the containment, collection, and return of leachate back through the landfill media. Many confirmed advantages are attributed to this operational method. It has been proven that leachate recirculating through sanitary landfills treats leachate partially and augments the stabilization rate of organic building blocks within the landfill (Christensen et al. 2005).

Leachate recirculating essentially converts the landfill into an anaerobic bioreactor by providing the contact time, nutrients, and substrates necessary for efficient change and degradation of refuse. Landfills utilizing recirculating generally exhibit higher and more efficient dilution of toxic substances and a much lower risk of harmful public health and environmental impacts (Christensen et al. 2005; Renou et al. 2008).

5.3.2 Recirculation in Landfills

Recycling leachate in MSW landfills can provide:

- Means of disposal (not only short term as it percolates through the waste but also by allowing the waste to absorb (soak-up) the leachate);
- Enhancement of the rate of landfill fixation (encouraging both the onset of fermentation leading to gas) and confined long term settlement; and
- Increased gas yields.

Leachate recycling would also appear to be a positive measure as the alternative (that is the absence of moisture within a landfill) is considered. Modern lined and capped containment landfill practice is often referred to as "dry-tomb landfilling", which will postpone the onset of emissions, rather than prevent them. If the waste is too dry it will never degenerate. If degeneration does not take place or a geological (or other) vent disrupts the lining, groundwater pollution will take place from landfill leachate. In reality, groundwater pollution will still happen unless decomposition and "flushing" has occurred when the containment fractures, but at least encouraging decomposition is a start. Even though leachate recycling has been gaining recognition worldwide, the merits of recycling MSW leachate are controversial.

5.3.3 Recirculating in Open Dumps

Recirculating of leachate in open dumps can also be proposed as the following:

Leachate recirculating can provide balance moisture during dry weather when leachate which would otherwise escape can be soaked back into the waste.

By augmenting the wetting of the waste, stabilization will be improved and if landfill gas can be amassed, the quantity of gas and the early payback potential to recoup the investment will be maximized. Making use of gas is another benefit to the local community as it produces bio-fuel energy that can be linked to the local power grid; and.

After initial fermentation/acetogenesis the recirculated leachate will be easier to treat aerobically.

Other relentless risks from recirculating arise if the level of leachate in the landfill is not carefully monitored and controlled. If leachate levels rise in the waste, breakouts may rapidly develop uncontrollably and bring about surface water pollution, however, worse can also occur. A number of landfills have suffered from the break-down of sloping faces for which the presence of high leachate levels has been a major if not the primary contributor.

One of the key goals of leachate recirculating is to optimize the water content in order to accelerate waste degradation. In the same way, the liquid flow is able to dilute the eventual presence of inhibitors and provides nutrients for biological degradation enhancement.

Beneficial effects on waste degradation and, in consequence, on biogas production, leachate organic load reduction, and waste settlements. Nevertheless, at large scale, optimization of water distribution and quantification of effects put still an important challenge forward.

5.3.4 Benefits of Leachate Recirculating

Leachate recirculating in MSW landfills offers these key benefits:

1. Reduction in leachate treatment and disposal costs;
2. accelerated decomposition and settlement of waste resulting in a gain in airspace;
3. Acceleration in gas production; and
4. Potential reduction in the post-closure care period and associated costs.

The most common ways for long-term leachate recirculating in MSW landfills include vertical injection wells and horizontal renches. Both of these methods end in the non-uniform distribution of leachate. In addition, the amount of leachate that can be recirculated by these methods is not enough to get rid of all leachate usually generated by landfills located in humid regions. Non-uniform distribution of leachate leads to uneven landfill settlement and as a result higher maintenance costs.

There are several methods of leachate recirculating to be applied into landfill such as:

1. Direct application of waste during disposal: The problems with this method
   include odor problems, health risks due to exposure, adjacency to landfill equipment and machinery, and off-site migration due to drift. This method also requires
   a leachate storage facility for periods such as high winds, rainfall, and landfill
   shutdowns when the leachate cannot be used.
2. Spray Irrigation of landfill surface: Here leachate is applied to the landfill surface
   in the same method that irrigation water is applied to crops. This method is
   beneficial because it allows the leachate to be used to a larger portion of the
   landfill, and also because the leachate volume is reduced due to evaporation.
3. Surface application: This is achieved through spreading the leachate. The ponds
   are generally formed in landfill areas that have been isolated with soil berms or
   within excavated sites in the solid waste. The disadvantages of these methods
   include an increasing amount of required land territory and monitoring of the
   ponds to detect seepage, leaks, and breaks that would make it possible for leachate
   to escape directly or with stormwater runoff.
4. Subsurface application: This is earned through placing either vertical recharge
   wells or horizontal drain fields within the solid waste.

5.3.5   Leachate Recirculating Stratagems

Several stratagems of leachate recirculating were tried to achieve the effective distribution of leachate. Originally, it was planned to spray leachate on the operating
face and other areas using spray headers. Spraying on the working face using a spray
nozzle also was tried which allowed for pliability in operation but was labor-intensive
and burdensome. Spraying also caused odor problems to landfill operators and equipment. The next stratagem tried was to excavate small pits in the waste and fill them
with leachate utilizing a spray header. Due to the shallow depth of the landfill, waste
had limited absorption capacity and consequently, the technique was abandoned.

To increase recirculating volumes, another stratagem was tried incorporating
trenches. Trenches were excavated on the completed parts of the landfill and filled
with leachate. The absorption capacity of the trenches varied and resulted in leachate
outbreaks in some parts of the landfill. Leachate outbreaks continued to occur and
coincided with periods of peak infiltration and recirculating. The trench method was
modified by filling the trench with auto-shredding derived waste or baled fiberglass
wastes. These materials acted as wicks and transmitted leachate to a larger area of
the refuse thereby increasing the permissible recirculating volumes and permitting
longer use of trenches. A combination of these techniques also was utilized.

5.3.5.1   Landfill Operation Strategies

Current regulations for MSW landfills emphasize on the landfill operation strategy
of containing and removing leachate before it enters the environment and limiting

the original generation of the leachate (Reinhart and Townsend 1997). The presence and movement of moisture in landfilled waste are one of the most important factors controlling waste degradation and landfill stabilization. Minimizing the amount of water degradation, landfill stabilization of the waste will be minimized too. This strategy has been termed the "dry tomb" approach; an alternate strategy, actively persuading the stabilization of the landfill through technologies such as leachate recirculating.

5.3.5.2 Landfills as Bioreactors

Under proper conditions, the rate of MSW biodegradation can be stimulated and enhanced. Environmental conditions that affect biodegradation most significantly include pH, temperature, nutrients, absence of toxins, moisture content, particle size, and oxidation–reduction potential. One of the most critical parameters influencing MSW biodegradation has been found to be moisture content. Moisture content can be most practically controlled via leachate recirculating. Leachate recirculating provides a means of optimizing environmental situations within the landfill to provide enhanced stabilization of landfill quantities as well as treatment of moisture moving through the fill (Reinhart and Townsend 1997; Townsend 2018).

The numerous advantages of leachate recirculating include distribution of nutrients and enzymes, pH buffering, dilution of inhibitory compounds, recycling and distribution of methanogens, liquid storage, and evaporation opportunities at low additional construction and operating cost. It has been suggested that leachate recirculating can reduce the time required for landfill stabilization from several decades to two or three years.

5.3.5.3 Bioreactor Landfill

Bioreactor landfill is a sanitary landfill that uses enhanced microbiological processes to transform and stabilize the readily and moderately decomposable organic waste constituents within 5 to 10 years of bioreactor process execution. The bioreactor landfill considerably increases the extent of organic waste decomposition, conversion rates and process effectiveness over what would otherwise occur within the landfill. Stabilization means that the environmental performance measurement parameters (landfill gas composition and generation rate and leachate constituent concentrations) remain at steady levels, and should not increase in the event of any partial containment system failures beyond 5 to 10 years of bioreactor process implementation. The bioreactor landfill requires certain specific management activities and operational modifications to improve microbial decomposition processes. The single most significant and cost-effective method is liquid addition and management (Qasim and Chiang 1994; Renou et al. 2008; Townsend 2018).

Numerous benefits can be derived from the bioreactor landfill as follows:

(a) Fast organic waste conversion and Stabilization

- Fast settlement: volume reduced and stabilized within bioreactor process implementation
- Increased gas unit yield: total yield and flow rate almost the entire rapid and moderately decomposable organic constituents will be degraded.
- Elevated leachate quality: stabilizes within 3 to 10 years after closure.
- Early land uses possible following closure.

(b) Maximizing of landfill gas capture for energy recovery projects

- Significant increase in total gas available for energy use, which provides
- entrepreneurial opportunities
- Potential increase in total landfill gas extraction efficiency (enabled over a shorter generation period)
- Increased greenhouse gas reduction from lessened emissions
- Increase in fossil fuel offsets due to increased gas energy sales
- Assistance in defraying landfill gas non-funded environmental costs
- The important economy of scale advantage due to high generation rate over relatively Short time.

(c) Increased landfill space capacity reuse due to rapid settlement during the operational time period.

- Increase in the amount of waste that can be placed into the permitted landfill airspace (Effective density increase.)
- Extension of landfill life through additional waste placement
- Postponed capital and financing costs needed to locate, permit and construct.
- Replacement landfill results in the capital and interest savings
- Significant increase in realized waste disposal revenues

(d) Improved leachate treatment and Storage

- Low cost partial or complete treatment; significant biological and chemical transformation of both organic and inorganic constituents, although mostly relevant to the organic constituents.
- Reintroduction of all leachate over most of the operational and post-closure care period significantly reduces leachate disposal costs.
- Absorption of leachate within landfill available up to field capacity

(e) A cut in post-closure care, maintenance, and Risk

- Rapid waste stabilization (within 5 to 10 years) minimizes environmental risk and liability due to settlement, leachate, and gas.
- Landfill operation and maintenance activities are considerably reduced
- Landfill monitoring activities can be reduced
- A cut in financial package requirement

In case of partial liner failure, there should be no risk of increased gas generation, worsening leachate quality, increased settlement rate or magnitude; another major benefit of bioreactors may come from greenhouse gas abatement. Bioreactors can generally rapidly complete methane generation while attaining maximum yield. This can be combined with nearly complete capture of generated gas using the bioreactor landfill in combination with a landfill gas energy project.

The stabilization of MSW goes on in five sequential phases is described in the subsequent sections. The rate and characteristics of leachate production and landfill gas generation from landfills are varying in different phases. These variations can be used for monitoring the stabilization of the MSW landfill. Five phases of MSW decomposition and stabilization are as follows:

Phase I: initial adjustment

This phase is related to the initial placement of MSW and the accumulation of moisture within the landfill. In this stage, biological degeneration occurs under aerobic conditions; in which oxygen is present in the void spaces of MSW. Microorganisms are provided from soil material or other sources like leachate recirculating, sludge. Moisture content is entered with incoming MSW to landfill, soil material covers and rainfall.

Phase II: transition

This phase triggers the transformation from aerobic to the anaerobic conditions because of oxygen depletion within the landfill. When the landfill situation is anaerobic, nitrate and sulfate are the electron acceptors in biological conversion reactions and will be reduced to nitrogen and hydrogen sulfide gas, and also the displacement of oxygen by carbon dioxide occurs. In this phase, the pH of the leachate starts dropping due to the presence of organic acids and the effect of the elevated carbon dioxide. By the end of this phase, chemical oxygen demand (COD) and volatile organic acids (VOA) or volatile fatty acids (VFA) can be detected in the leachate.

Phase III: acid formation

The continuous hydrolysis (solubilization) of solid waste and biological activities of microorganisms turn biodegradable organic content at high concentrations into intermediate volatile fatty acids. Decreasing pH values is often observed, followed by metal species mobilization. Rapid consumption of substrate and nutrients occurred in this phase.

Phase IV: methane fermentation

Intermediate acids from phase III are consumed by methanogenic bacteria and converted to methane and carbon dioxide. Sulfate and nitrate are reduced to sulfide and ammonia, respectively. The pH values rise by the bicarbonate buffering system, this condition will support the growth of methanogenic bacteria. Heavy metals are emitted by compellation and precipitation.

Phase V: maturation

In this phase, nutrients and available substrate become limiting, and biological activities slow down. Gas production drops dramatically and leachate strength stays steady at lower concentrations. The reappearance of oxygen and oxidized sorts may be observed slowly.

During the maturation phase, the leachate will often contain humic acid and folic acid, which are difficult to process further biologically.

5.4   Mathematical Modeling of Leachate Recirculating

As far as many parameters affect moisture routing through a landfill, a mathematical model of the recirculating landfill was employed to consider the impact of these parameters on design and operational needs more manifestly. The U.S, Geological Survey (USGS) model for Saturated and Unsaturated Flow and Transport, SUTRA 82, 83 was used to model the recirculating landfill. SUTRA uses a two-dimensional hybrid finite element and an integrated finite difference method to approximate the governing equations of flow and transport. SUTRA is capable of performing steady-state and non-steady-state simulations. Modeling proceeded in two phases: steady-state modeling to perform an initial screening of SUTRA capabilities and to identify further data needs and transient modeling following extensive modifications to SUTRA. Two recirculating methodologies were selected for simulation, the vertical leachate infiltration well, and the horizontal infiltration trench. These methodologies were selected based on their vast use and their compatibility with the final closure of the landfill. Landfill depths of 6 and 18 m (20 and 60 feet) were modeled. The preliminary inputs to SUTRA are the physical characteristics of both the solid matrix and fluid, porosity, permeability, dispersivity, and unsaturated flow characteristics. Porosity is input on a node-wise basis while permeability and dispersivity are input by element. The basis of the SUTRA simulation is a mesh of nodes in Cartesian coordinates which are then connected to form quadrilateral elements. The output from the model provides a degree of saturation (volume of water/volume of voids), fluid mass budgets, and depth of the head on the landfill liner as a function of the rate of leachate introduction and the site of the recirculating device(s) (Renou et al. 2008; McCreanor and Reinhart 2000; 1996; Mccreanor 1999).

The power equations presented by Korfiatis et al. were used to model the unsaturated characteristics of the waste matrix. Korfiatis et al. determined the saturated suction head to be 6 cm of Water for municipal solid waste. The Brooks and Corey equations with appropriate parameters were used to model the sand and gravel components of the model.

$$k_r = \left(\frac{h}{h_s}\right)^{-2.75}$$

$$\theta = \left(\frac{h}{h_s}\right)^4$$

where

    k_r = relative hydraulic conductivity, (unitless).
    θ = volumetric moisture content, wet basis, (V/V).
    h_s = saturation suction pressure, (ML-1 T-2).
    h = suction pressure, (ML-1 T-2).

5.4.1 Trench Modeling

The modeling of the infiltration trench consisted of placing discrete fluid sources directly within the waste matrix. The gravel/tire chip and geotextile materials commonly used within the trench were not modeled due to dimensionality limitations and also in order to simplify the construction of the required input files. The recirculating rates modeled were 2.0, 4.0, 6.0, and 8.0 m³/day/m of a trench which bracket reported operating ranges of horizontal infiltration trenches.55 Transitory simulations of single and multiple horizontal injection lines have been conducted for a saturated hydraulic conductivity of 10–3 cm/s and a landfill depth of 15 m.

5.4.2 Vertical Infiltration Well Modeling

The modeling of the vertical infiltration well-required usage of a radial coordinate system and placed a series of fluid sources vertically from a level of 2 m to 13 m, discharging directly into the waste. Simulation of recirculating rates of 0.20, 0.40, and 0.80 m³/day at hydraulic conductivity of 10–3 cm/s and a landfill depth of 15 m were conducted. It was also indispensable to model the effect of increasing fluid pressure with depth within the well on discharge rate.

5.4.3 Model Results

Degree of saturation (volume of water divided by volume of pore spaces) iso-clines for the horizontal trench recirculating leachate at rates of 2.0 and 8.0 m³/day/m of the trench are examined. The influence distance was defined as the lateral distance from the trench to which the saturation had been increased above the initial condition. The amounts of saturation iso-clines suggest that flow rates of 6.0 and 8.0 m³/day/m trench results in the upward propagation of a saturated front and artesian conditions at the landfill surface.

5.5   Leachate Recirculating Field Testing

Leachate recirculating is a popular and economic mehod can be use in developing and arid countries like Iran. In this part, leachate recirculating field method discussed. Mathematical modeling is an extremely useful tool that allows parameters to be varied without the expense of physical tests. However, without field verification the validity of model results is questionable.

5.5.1   Leachate Treatment Implications

Ideally, the bioreactor landfill should be operated to minimize offsite management of leachate; however, eventually, the treatment of leachate will be necessary. Following extended recirculating, the leachate will be largely devoid of biodegradable organic material and will contain recalcitrant organic and inorganic amalgams such as ammonia, chloride, iron, and manganese. Treatment needs depend upon the final disposition of the leachate. Final disposal of leachate may be accomplished through co- disposal at a publicly owned treatment works (POTW) or through onsite treatment and direct discharge to a receiving body of water, deep well injection, land usage, or via natural or mechanical evaporation.

Leachate treatment can be challenging because of low biodegradable organic strength, irregular production rates and composition, and low phosphorous content (if biological treatment is desired). Due to the nature of leachate, physical/chemical treatment processes such as ion exchange, reverse osmosis, chemical precipitation/filtration, and carbon adsorption are the most probable options.

5.5.2   Leachate Quantities

Recirculated leachate represents an increasing percentage of generated flows (asymptotically approaching 100%) as recirculating rates increase. With the current landfill capping practices, recirculated leachate volumes will become especially dominant after landfill closure. As with conventional landfills, leachate generation is a function of climate and site characteristics, as well as leachate recirculating amounts.

Offsite disposal of leachate ranged from 0 to 59% of generated leachate. Data suggested that, unlike conventional landfills, the amount of leachate requiring offsite management at recirculating landfills was a function of both the volume of leachate generated and the available onsite storage. At sites where large storage volumes were provided relative to the size of the landfill cell, offsite management of leachate was minimized (frequently no offsite management was required for long periods of time). It was also observed that sites with relatively little storage were compelled to recirculate leachate at much higher amounts than those with large storage volumes.

5.5.3 Gas Production

While gas production is relatively easy to determine from laboratory lysimeters, full-scale measurement of gas emissions from active sites is more difficult to achieve. Limited data suggest that, as in lysimeters, gas production significantly is enhanced at large-scale landfills as a result of both accelerated gas production rates as well as the return of organic material in the leachate to the landfill for conversion to gas (as opposed to washout in conventional landfills). Gas emission measurements were made by researchers at a recirculating landfill using a patented device, the Flux Tube, a variation on the flux chamber used to measure surface emissions. These tests disclosed a doubling of gas production rates from waste located in wet areas of the partially recirculating landfill relative to comparably aged waste in dry areas 94. This reality was corroborated by measurements of biological methane potential (BMP) from samples obtained in wet and dry areas of the same landfill. 85 A 50% decrease in BMP was measured in wet examples (46% wet basis) over a one-year period.

5.6 Leachate Recirculating System Design

According to the evaluation of available data, the most practical and efficient recirculating methodology utilizes horizontal devices, vertical devices, or a combination of horizontal and vertical systems. The design criterion for the placement of reintroduction devices is scarce and typically based on the previous experience. Other issues remain uncertain in designing for full-scale leachate recirculating including the determination of the area of influence of recirculating devices, the effect of leachate recirculating on leachate collection systems, and appropriate recirculating flow rates.

5.6.1 Horizontal Trenches

The results of the modeling efforts for horizontal trench are that the distance impact increases as the flow rate rises. The simulation was conducted assuming a waste hydraulic conductivity of 10–3 cm/sec. At lower hydraulic conductivities, it is expected that greater horizontal spreading will result, however, the downward movement will be impaired. As a result, horizontal spacing can be increased; however, vertical spacing must be reduced. Influence distance also appears to increase when intermittent leachate introduction is practiced. The influence distances should be used as a guideline, particularly when placing horizontal trenches near the landfill surface and boundaries.

5.6.2   Vertical Recharge Wells

Saturation profiles for the vertical well suggest that the leachate will initially show preferential flow vertically along the well surface. Such flow may contribute to the localized subsidence around the wells at full-scale sites. Higher saturation (greater than 0.6) initially develops along the well surface and slowly begins to broadcast laterally and vertically as leachate attempts to percolate downward more quickly than it can be conveyed by the waste matrix. Modeling found that the impact area of a vertical well was a function of the rate of flow, as the horizontal trench. However, in these cases, small increases in inflow resulted in large augments in the impact area. Vertical well spacing is conventionally 35 to 100 m (118 to 333 ft). Al-Yousfr78 proposed a simplified estimate of vertical well influence radius based on the relative hydraulic conductivity of- the well filling (usually gravel) and the waste, as shown in Eq. 7.4.

$$R = \frac{rK_w}{K_r}$$

where: R = radius of influence zone, (L).
   r = radius of the recharge well, (L).
   K_w = permeability of media surrounding well, (L/T).
   K_r = permeability of refuse, (L/T).

5.6.3   Design Approach

The design of a recirculating system should include more than one recirculating system. Initial wetting with leachate as the waste is placed is also recommended. Once sufficient waste is in place (3 to 6 m [10 to 18ft]), horizontal trenches and vertical wells can be utilized. Trench spacing can be determined using the following procedure:

1.   Determine the desired volume to be recirculated, using hydrologic modeling results, i.e., HELP.
2.   Determine dimensions of the landfill
3.   Select the length of trenches.
4.   Determine the flow rate per length of the trench.
5.   Determine the spacing of the trench.

   To the trench during portions of the day and rotated from trench to trench to increase flow to each trench. Low flow rates will end in incomplete wetting of the landfill. If the flow is above the curve, the spacing of the trenches should be reduced. Note that the influence distance should be doubled to determine trench spacing requirements.

References

Amokrane A, Comel C, Veron J (1997) Landfill leachates pretreatment by coagulation-flocculation. Water Res 31(11):2775–2782

Aziz HA, Amr SA (Eds) (2016) Control and treatment of landfill leachate for sanitary waste disposal. Hershey: IGI Global

Campos JC, Moura D, Costa AP, Yokoyama L, Araujo FVDF, Cammarota MC, Cardillo L (2013) Evaluation of pH, alkalinity and temperature during air stripping process for ammonia removal from landfill leachate. J Env Sci Health, Part A 48(9):1105–1113

Chian ES, Dewalle FB (1976) Sanitary landfill leachates and their treatment. J Env Eng Div 102(2):411–431

Christensen TH, Cossu R, Stegmann R (Eds.) (2005) Landfilling of waste: leachate. CRC Press

Delarestaghi RM, Ghasemzadeh R, Mirani M, Yaghoubzadeh P (2018) The comparison between different waste management methods of Tabas city with life cycle assessment assessment. J Env Sci Studies 3(3):782–793

Diaz LF (1999) Book reviews: landfill bioreactor design & operation by Debra R. Reinhart & Timothy G. Townsend Boca Raton, Florida: CRC Press/Lewis Publishers/St Lucie Press. Waste Manag Res 17(3):246–247

Di Palma L, Ferrantelli P, Merli C, Petrucci E (2002) Treatment of industrial landfill leachate by means of evaporation and reverse osmosis. Waste Manag 22(8):951–955

Foul AA, Aziz HA, Isa MH, Hung YT (2009) Primary treatment of anaerobic landfill leachate using activated carbon and limestone: batch and column studies. Int J Env Waste Manag 4(3–4):282–298

Ghasemzade R, Pazoki M (2017) Estimation and modeling of gas emissions in municipal landfill (Case study: Landfill of Jiroft City). Pollution 3(4):689–700

Ghasemzadeh R, Pazoki M, Hoveidi H, Heydari R (2017) Effect of temperature on hydrothermal gasification of paper mill waste, case study: the paper mill in North of Iran. J Env Studies 43(1):59–71

Gulsen H, Turan M (2004) Treatment of sanitary landfill leachate using a combined anaerobic fluidized bed reactor and Fenton's oxidation. Env Eng Sci 21(5):627–636

Henry JG, Prasad D, Young H (1987) Removal of organics from leachates by anaerobic filter. Water Res 21(11):1395–1399

Lema JM, Mendez R, Blazquez R (1988) Characteristics of landfill leachates and alternatives for their treatment: a review. Water, Air, and Soil Pollution 40(3–4):223–250

Loizidou M, Papadopoulos A, Kapetanios EG (1993) Application of chemical oxidation for the treatment of refractory substances in leachates. J Env Sci Health Part A 28(2):385–394

Madu JI (2008) New leachate treatment methods. Department of Chemical Engineering, Lund University, Sweden

McArdle JL, Arozarena MM, Gallagher WE (1988) Treatment of hazardous waste leachate: unit operations and costs

Mccreanor PT (1999) Landfill leachate recirculation systems: mathematical modeling and validation

Mccreanor PT, Reinhart DR (1996) Hydrodynamic modeling of leachate recirculating landfills. Water Sci Technol 34(7–8):463–470

McCreanor PT, Reinhart DR (2000) Mathematical modeling of leachate routing in a leachate recirculating landfill. Water Res 34(4):1285–1295

Mulamoottil G, McBean EA, Rovers F (2018) Constructed wetlands for the treatment of landfill leachates. Routledge

Pazoki M, Abdoli MA, Ghasemzade R, Dalaei P, Ahmadi Pari M (2016) Comparative evaluation of poly urethane and poly vinyl chloride in lining concrete sewer pipes for preventing biological corrosion. Int J Env Res 10(2):305–312

Pazoki M, Abdoli MA, Karbassi A, Mehrdadi N, Yaghmaeian K (2014) Attenuation of municipal landfill leachate through land treatment. J Env Health Sci Eng 12(1):12

Pazoki M, Abdoli M, Karbasi A, Mehrdadi N, Yaghmaeian K, Salajegheh P (2012) Removal of nitrogen and phosphorous from municipal landfill leachate through land treatment. World Appl Sci J 20(4):512–519

Pazoki M, Delarestaghi RM, Rezvanian MR, Ghasemzade R, Dalaei P (2015) Gas production potential in the landfill of Tehran by landfill methane outreach program. Jundishapur J Health Sci 7(4)

Pazoki M, Ghasemzadeh R, Yavari M, Abdoli M (2018) Analysis of photocatalyst degradation of erythromycin with titanium dioxide nanoparticle modified by silver. Nashrieh Shimi va Mohandesi Shimi Iran 37(1):63–72

Pazoki M, Ghasemzade R, Ziaee P (2017) Simulation of municipal landfill leachate movement in soil by HYDRUS-1D model. Adv Env Technol 3(3):177–184

Pazoki M, Pari MA, Dalaei P, Ghasemzadeh R (2015) Environmental impact assessment of a water transfer project. Jundishapur J Health Sci 7(3)

Qasim SR, Chiang W (1994) Sanitary landfill leachate: generation, control and treatment. CRC Press

Reinhart DR, Townsend TG (1997) Landfill bioreactor design & operation. CRC press

Renou S, Givaudan JG, Poulain S, Dirassouyan F, Moulin P (2008) Landfill leachate treatment: review and opportunity. J Hazardous Mater 150(3):468–493

Shayesteh AA, Koohshekan O, Khadivpour F, Kian M, Ghasemzadeh R, Pazoki M (2020) Industrial waste management using the rapid impact assessment matrix method for an industrial park. Global J Env Sci Manag 6(2):261–274

Timur H, Özturk I (1999) Anaerobic sequencing batch reactor treatment of landfill leachate. Water Res 33(15):3225–3230

Townsend TG (2018) Landfill bioreactor design & operation. Routledge

Wiszniowski J, Robert D, Surmacz-Gorska J, Miksch K, Weber JV (2006) Landfill leachate treatment methods: a review. Env Chem Lett 4(1):51–61

Zhang T, Ding L, Ren H (2009) Pretreatment of ammonium removal from landfill leachate by chemical precipitation. J Hazardous Mater 166(2–3):911–915

Chapter 6
Natural Treatment Systems

6.1 Introduction

Using land for getting rid of wastes has been the most prevailing method for a far long time. The soil has been impeccable in terms of absorbing and stabilizing body wastes generated by a sparse population of nomadic people or animals, thus no problem was posed then (Ghasemzadeh et al. 2017; Shayesteh et al. 2020). The early records of problematic conditions due to dense population date back to biblical times when the need for the establishment of an organized management system for waste disposal was first felt. As more people were adopted centralized habitation in cities, a novel technology was devised to bring waste disposal under control. The first land management system in the form of sewage irrigation was initiated in Bunzlau, Germany in 1531, and endured for over 300 years. Another similar system was put into operation in a region near Edinburgh, Scotland in about 1650. Usability of sewage as a type of fertilizer for producing agricultural crops was formally admitted at that time (Ghasemzade and Pazoki 2017; Pazoki et al. 2017, 2018).

Land treatment is a process that is designed to minimize the toxicity of the soil by treating the hazardous components of the sewage that is applied to the land. The most common processes that are mainly used for this purpose include slow rate (SR), overland flow (OF), and rapid infiltration (RI). The matrix of soil–plant-water is treated by the same physical, chemical, and biological mechanisms in natural ways in all these processes. The treatment in both SR and SAT processes is accomplished on the soil matrix once the sewage has been infiltrated. They only differ in the rate by which the wastewater is entered into the site. In the case of the process, the soil and vegetation on the surface are used for treatment. Permeation is limited in this process, so the outflow is gathered after passing through the treating stage at the bottom of the slope as surface runoff (Renou et al. 2008; Qasim and Chiang 1994).

Since both construction and application of such systems are desirable in terms of cost-effectiveness, they appear as the optimum alternatives thus are widespread in small urban or rural communities.

© Springer Nature Switzerland AG 2020
M. Pazoki and R. Ghasemzadeh, *Municipal Landfill Leachate Management*,
Environmental Science and Engineering,
https://doi.org/10.1007/978-3-030-50212-6_6

The sewage generated by resident households in these regions can be used for irrigating certain species of plants with a quick growth rate, thus it is an effective approach to treat wastewater and supplying necessary nourishing substances to grow plants. Although sewage can be useful in supplying irrigating water for agricultural crops, it makes these products susceptible to harmful insects and disease-causing organisms. Nevertheless, this approach to fertilization of soil through irrigating trees with wastewater is both cost-effective and environment-friendly. Some species of trees such as Poplar and Salix can even take much more advantage of the nutrients that naturally exist in wastewater because growing seasons are longer in these trees, and they have deeper and long-lasting roots than annual plants. On the other hand, the rate of evapotranspiration in such trees is so high that it enables LTS treatment to yield higher efficiency (Qasim and Chiang 1994; Mulamoottil et al. 2018; Christensen et al. 2005).

Certain factors including the mode of sewage application, chemical specifications of wastewater, and the local soil profile determine the technical design of the land treatment system. Moreover, some parameters such as dissolved salts, suspending solids, nutrients like nitrogen and phosphorus, organic matter, cations like sodium and magnesium, and toxic substances have to be taken into account. Depth of the soil layers, depth of groundwater table, the angle of slope in the area, and the soil permeability are regarded as significant conditions of the site. The modes of sewage treatment using land usually fall into the following categorizations:

1. Slow Rate (SR) method
2. Rapid Infiltration (RI)
3. Overland Flow (OF).

6.1.1 Slow Rate Method

This method is perhaps the most conventional and the most prevalent technique in the field of land treatment of wastewater. The fundamental process has passed several developmental stages from sewage farming in Europe around the sixteenth century to a more advanced treatment system in England in the 1860s. Some slow-rate systems had been established in the US by the 1880s. The slow-rate land treatment system was rated as the most prevalent method according to an 1899 survey that was conducted on a totally of 143 various plants for sewage treatment. In the middle of the 1960s, slow-rate land treatment was appreciated again at Penn State. Research and development in the field of land treatment of sewage absorbed interests of the US Environmental Protection Agency (USEPA) and the US Corps of Engineers to invest much more time and energy on this domain. A wide range of studies with long-term impacts had been carried out by the late 1970s. Today, one large SR system covers 4605 acres of land across a sprinkler-irrigated forest in Dalton, Georgia (Christensen et al. 2005; Crites et al. 2014).

6.1.2 Rapid Infiltration

Rapid infiltration that is also called soil aquifer treatment is another variety of land treatments in which the wastewater is subjected to treatment when it penetrates into the soil matrix. Physical, chemical, and biological measures taken to treat the wastewater are not restricted to the surface but they extend to lower layers as the effluent penetrates into the vadose zone and groundwater. This approach is usually adopted when the soil is more permeable to a considerable depth. The operations that are applied in this process are usually sporadic and extend to shallow basins within the penetration scope. Since there is no alternate oxidation/reduction condition in this process, the output from treatment is not so perfect in spite of continuous flooding or ponding operations. Plants cannot be taken into consideration in the RI systems because the feeding rate is so high that it makes nitrogen uptake less effective. However, considering the critical effects of vegetation on stabilization of the soil and keeping the infiltration rate at a high level in certain conditions, local plants play an essential role in treatment systems.

6.1.3 Overland Flow

This method has been specially designed for the lands with soils of clay type into which wastewater penetrates slowly. Treating operation in this process is accomplished as effluent streams downward through vegetated and even-graded soil with a gentle slope ranging from 2 to 8%. Once the treatment has been completed, the outflow of runoff is gathered at the bottom of the slope. This process was initially innovated by Campbell Soup Company in the US, which was first established at Napoleon, Ohio in 1954 and then at Paris, Texas. A related study about the OF process was carried out on urban sewage produced from Ada, Oklahoma and Utica, Mississippi. Based on the findings from this and other relevant studies, more than 50 different OF systems were built for treating urban sewage (Crites et al. 2014).

6.2 Designing Method and Fundamental Technology

The concept of Limiting Design Parameter (LDP) plays a central role in designing all land-treatment systems, wetlands management, and other processes in a similar field. LDP is an essential factor based on which the design, allowable size, and acceptable load of a particular system are determined. A typical system will only succeed in taking all critical parameters into account when the initial design meets the LDP.

Table 6.1 Site characteristics for land treatment processes

Parameter	Slow rate (SR)	Rapid infiltration (RI)	Overland flow (OF)
Grade	20%, cultivated site 40%, uncultivated	Not critical	2 to 8% for final 40%, uncultivated slopes
Soil permeability Groundwater depth	Moderate 2–10 ft	Rapid 3 ft during application 5–10 ft during drying	Slow to none Not critical
Climate	Winter storage in cold climates	Not critical	Same as SR

The obtained results from the observations on treatment systems for urban sewage, particularly those that substantially depend on infiltration like SR and RI, have indicated that LDP is mainly associated with either the hydraulic capacity of the considered soil or possibility of minimizing nitrogen to a definite level. Between these two parameters, the one will determine the design meeting LDP that necessitates the largest area of treatment. Once the design is complete, the remaining operational requirements should be met subsequently. LDP in the case of overland flow system as a disposal arrangement will depend on the site-specific disposal restrictions, then the design will be determined according to the parameter that necessitates the largest area of treatment.

As already mentioned, sewage treatment involves three main processes: slow rate (SR), rapid filtration (RI), and overland flow (OF). They refer to both the water displacement rate and the stream path for each process. However, this field is not only limited to these processes, but there are also some additional criteria for integrated systems i.e. wetland and alternative technology, on-site and small-scale systems, and certain criteria with regard to applying biosolids for land treatment (Bhargava et al. 2016).

Table 6.1 summarizes the main specifications of an ideal site of all three processes. However, these properties are only general rules of thumb and may vary based on the particular circumstances and needs.

6.2.1 Characteristics of Design

General criteria of design for these treatment processes do not compare. The recommended values have been deduced from successful applications in various regions throughout the United States.

Table 6.2 Resulted in average values from treatment that will occur within the immediate plant-soil

Parameter	Slow rate (SR)	Rapid infiltration (RI)	Overland flow (OF)
Application method	Sprinkler or surface	Usually surface	Sprinkler or surface
Annual loading, ft	2–20	20–400	10–70
Treatment area for 1 mgd, acres	60–700	7–60	15–110
Weekly application, in	0.5–4	4–96	2.5–16
Minimum preliminary treatment	Primary	Primary	Grit removal and comminution
Need for vegetation	Required	Grass sometimes used	Water-tolerant grasses

Table 6.1 indicates the anticipated quality of effluent running off each land treatment process with regard to the most frequently-concerned parameters.

The given data in Table 6.2 are average values associated with plant-soil matrix regardless of mixing, dispersion, or dilution of the groundwater or further permeation into lower layers of the soil. As wastewater penetrates into the farther levels of the soil, phosphorus may be reduced to a level that is at least one order above the magnitude as typical for RI systems.

6.2.2 Expected Performance

Table 6.3 summarizes the anticipated optimum quality of the effluent resultant from the three major land treatment processes with regard to the most frequently-concerned

Table 6.3 Expected effluent water quality from land treatment

Parameter	Slow rate (SR)	Rapid infiltration (RI)	Overland flow (OF)
BOD5	<2	5	10
TSS	<1	2	10
NH3/NH4 (as N)	<0.5	0.5	<4
Total N	3	10	5
Total P	<0.1	1	4
Fecal coli (number/100 mL)	0	10	200+

parameters. The given data in Table 6.2 are average values associated with plant-soil matrix regardless of mixing, dispersion, or dilution of the groundwater or further permeation into lower layers of the soil. As wastewater penetrates into the farther levels of the soil, phosphorus may be reduced to a level that is at least one order above the magnitude as typical for RI systems.

6.2.3   Relevant Parameters for Leachate

Leachate refers to the liquid that drains from special landfill sites allocated for municipal solid waste (MSW) that is highly polluted and is considered as one of the difficult types of wastewater to cope with. Leachate is produced when precipitation percolates into an open landfill or through the cap of a closed landfill. Leachate is usually characterized by a high concentration of organic matter (biodegradable and non-biodegradable), ammonia nitrogen, heavy metals, and chlorinated organic and inorganic salts. Properties of leachate may vary within a wide range that depends on the waste composition, amount of precipitation, hydrology of the site, compression of the waste, design of the cover, routines used for sampling, and also how leachate correlates with the environment, design of landfill, and the applied operation.

Biological oxygen demand (BOD_5) and chemical oxygen demand (COD) are two major factors based on which the organic content of leachate is assessed. The concentrations of the substances that pollute leachate may vary over several orders of magnitude. The followings are some of the significant properties that characterize leachate.

6.2.3.1   pH

The pH level of leachate usually ranges from 4.5 to 9. Young landfill leachates often have pH below 6.5 while that is over 7.5 in the case of old leachates. High concentrations of volatile fatty acids (VFAs) account for lower pH during the early life of a landfill. Once the leachate has been stabilized, pH finds a slightly constant level with negligible variations. In this period, pH may vary between 7.5 and 9 which is comparable to the pH level typical of old landfills ranging from 7.46 to 8.61 and 7.3 to 8.8 respectively.

6.2.3.2   TDS

The major substances that are contained in TDS include inorganic salts and dissolved organics. TDS accounts for a significant factor that is often taken into consideration in giving official permissions for disposal of landfill leachate in many countries including the UK. TDS indicates the amount of mineralization; thus the higher concentration TDS has, the more physical and chemical changes it may induce upon the receiving water. As the value of TDS concentration augments, the amount of salinity increases, and it, in turn, leads to higher toxicity due to changes taking place in the ionic composition of water.

6.2.3.3 BOD$_5$ and COD

Higher levels of BOD$_5$ and COD are prominent indicators of leachate during the first acidogenic biodegradation phase. The most common characterizations of leachate from young landfills are high levels of BOD$_5$ (4000–13,000 mg/L) and COD (30,000–60,000 mg/L). Tatsi et al. claimed that BOD$_5$ in leachate from young landfills may reach as high as 81,000 mg/L. The obtained results from a case study conducted by Tatsi and Zouboulis in Thessaloniki Greater Area (Greece) demonstrated a much higher value of COD (70,900 mg/L). As the landfill gets older, the amounts of BOD$_5$ and COD often decrease gradually. COD usually varies from 5000–20,000 mg/L when leachate reaches a stabilized state. The most useful way for evaluating the current state of leachate is to estimate BOD$_5$/COD ration which usually ranges from 0.4 to 0.5 for young leachate. As leachate enters the methanogenic phase, certain bacteria such as *methanogenic archaea* decreases the organic strength of leachate. Therefore, the concentration of VFAs will decrease subsequently, and this makes the BOD$_5$/COD ratio reach less than 0.1.

6.2.3.4 Total Nitrogen

Ammonium accounts for the largest part of the total nitrogen. The release of soluble nitrogen into leachate takes a longer time compared with that of soluble organics. Consequently, as the landfill gets older, the concentration of ammonia nitrogen will undergo a considerable increase that is attributable to hydrolysis and fermentation of nitrogenous proportions of biodegradable substances found in the waste. Ammonia is highly stable under anaerobic conditions, so it is usually deemed to be an essential contaminant with long-term effects. The concentration of ammonia in leachates may vary from one landfill to another ranging widely from tens or hundreds of mg N_{NH4}/L to a couple of thousands (2000 mg N_{NH4}/L) and to several thousand (>10,000 mg N_{NH4}/L). The average value of ammonia concentration in leachate may vary from 500 to 1500 mg/L following 3–8 years after disposal and usually remains unchanged after 50 years. Li and Zhao found a range of 3000–5000 mg/L for ammonia nitrogen in the case of stabilized leachates. A wide range of close analyses using bioassays and different test organisms such as *Salmo gairdnieri* and Oncorhynchus *nerka* have demonstrated the toxicity of ammonia nitrogen to living organisms. Similar studies have also approved that ammonia with high contents can stimulate algal growth and assist eutrophication because of the lower level of dissolved oxygen. Being a toxicant, it may disrupt the order of required treatment operations to be implemented biologically on the leachate.

6.2.3.5 Heavy Metals

Leachate from landfills typically contains a relatively low content of heavy metals. However, the low pH due to the generation of organic acids keeps the concentration of heavy metals at a rather higher level, since metals are more soluble during the initial stages. By gradual decrease of pH value in subsequent phases, the rate of metal solubility declines, thus the content of heavy metals will decrease accordingly. Lead is an exception to this rule since it can produce very heavy complexes with humic acids. In a study have done by Christensen et al. and Kjeldsen and Christophersen on 106 different landfills throughout Denmark, the researchers detected low concentrations of Cd (0.006 mg/L), Ni (0.13 mg/L), Zn (0.61 mg/L), Cu (0.07 mg/L), Pb (0.07 mg/L), and Cr (0.08 mg/L). Metals lose their toxicity in the presence of dissolved organic carbon (DOC) with high concentration since only free metals can appear as toxicants. Nevertheless, natural and synthetic complex ligands such as EDTA and humic materials can enhance the solubility and mobility of metals. Moreover, colloids have a strong attraction towards heavy metals, and a considerable but changing proportion of heavy metals interacts with colloidal matter. Baun and Christensen argue that free forms of metal ions contain less than 30%, usually less than 10%, of the total metal content, and the remaining can be found in colloidal or organic complexes. Jensen and Christensen concluded that 10–60% of Ni, 30–100% of Cu, and 0–95% of Zn are formed in colloidal portions. Furthermore, as a condition of leachate decreases and changes the ionic state of the metals i.e. Cr (VI) → Cr(III), and As (V) → As (III), the solubility of the metals will rise.

6.2.3.6 Phenol

Christensen et al. found phenolic compounds in landfills with a concentration ranging from 1 to 2100 μg/L. Phenolic compounds can easily degrade under aerobic conditions; however, there is no clear evidence showing their degradability under anaerobic conditions yet.

6.2.3.7 Chlorides

Deng and Englehardt examined several landfills and found chlorides with concentrations between 200 and 3000 mg/L in the case of relatively young landfills (1–2 years old); however, these values decrease to 100–400 when a landfill reaches the age of over 5–10 years. Bowman et al. examined leachate from a landfill in the Newington area in Sydney and measured the concentration of chlorides as high as 8000 mg/L. Since the soil is not able to weaken chlorides and they are active under any condition, a substantial concern has been focused on them as tracing constituents of leachate in association with groundwater.

6.2.3.8 Cyanide

Bagchi (1994), Tchobanoglouset et al. (1993) and Oweis and Khera (1990) estimated the contents of cyanide in leachate within a range from 0 to 6 mg/lit.

6.2.3.9 Total Coliform Bacteria

The obtained results from examinations carried out by Bagchi (1994), Tchobanoglous et al. (1993) and Oweis and Khera (1990) show that the total content of Coliform Bacteria in leachate varies from 0 to 100 mg/lit. The common characteristics of leachate identified by Bagchi (1994), Tchobanoglous et al. (1993) and Oweis and Khera (1990) are summarized in Table 6.3.

6.2.4 System Interactions

6.2.4.1 Biochemical Oxygen Demand

Biochemical oxygen demand (BOD_5) generally refers to the efficiency of land treatment approaches in terms of removing the biodegradable organics. Removal mechanisms include filtration, adsorption, and biological reduction and oxidation. Slow rate (SR) and rapid infiltration (RI) processes both react mostly across the ground surface or sub-surface soils where the microbial activity is at its highest intensity. In the case of overland flow (OF), relevant reactions fundamentally take place over the surface or within the mass of plant refuse and microbial substances.

In the case of OF systems, a considerable portion of particulate substances deposits when wastewater flows through a thin film streaming down the slope. Nevertheless, the condition is not the same for the elimination of algae because the confinement time on the slope is not long enough to enable complete removal through the physical deposition. The biological growths and sludge that are formed on the sloping part of an OF system, fundamentally account for the removal of contaminants.

The treatment process is fundamentally based on a biological mechanism that enables the BOD_5 removal capacity to be continuously renewable, provided that the loading rate and cycle meet the requirements of preservation or restoration of aerobic conditions in the system. In 1998, several pilot studies assessing soil columns showed that removal of BOD_5 to as low as background level is not dependent on the level of pretreatment, soil type, and infiltration rate. These findings are consistent with the data given in Table 6.4 (Bagchi 1994; Tchobanoglous et al. 1993; Oweis and Khera 1990; Pazoki et al. 2016). They also approved that BOD_5 removal in land treatment systems does not depend on high levels of pre-treatment.

Table 6.4 Typical constituents of leachate from MSW landfills

Parameter	Range (mg/l)		
Type	Parameter	Minimum	Maximum
Physical	pH	3.7	8.9
	Turbidity	30JTU	500JTU
	Conductivity	480 mho/cm	72,500 mho/cm
Inorganic	Total Suspended Solids	2	170,900
	Total Dissolved Solids	725	55,000
	Chloride	2	11,375
	Sulphate	0	1850
	Hardness	300	225,000
	Alkalinity	0	20,350
	Total Kjeldahl Nitrogen	2	3320
	Sodium	2	6010
	Potassium	0	3200
	Calcium	3	3000
	Magnesium	4	1500
	Lead	0	17.2
	Copper	0	9.0
	Arsenic	0	70.2
	Mercury	0	3.0
	Cyanide	0	6.0
Organic	COD	50	99,000
	TOC	0	45,000
	Acetone	170	11,000
	Benzene	2	410
	Toluene	2	1600
	Chloroform	2	1300
	Delta	0	5
	1,2 dichloroethane	0	11,000
	Methyl ethyl ketone	110	28,000
	Naphthalene	4	19
	Phenol	10	28,800
	Vinyl Chloride	0	100

(continued)

Table 6.4 (continued)

Parameter	Range (mg/l)		
Type	Parameter	Minimum	Maximum
Biological	BOD	0	195,000
	Total Coliform bacterial	0	100
	Fecal Coliform bacterial	0	10

6.2.4.2 Organic Loading

Comparing the given data in Table 6.5, we can easily find that land treatment systems possess a considerable capacity for treatment of the degradable organics that are typical characteristics of BOD_5. The effluents leaving both RI and SR systems are very similar with only a little difference with regard to organic loading that is about one order of magnitude higher for RI systems.

Five SR systems in Idaho were analyzed for wastewater from potato processing industry applying chemical oxygen demand (COD) loadings which varied between 40 and 280 lb/(acre.day) and revealed 98% removal following a 5-ft penetration into the soil. An OF treatment system dedicated to strong wastewater from snack food processing yielded favorable output in terms of BOD_5 loading rates within a range of 50 to 100 lb/(acre. day). Similar desirable results were observed in pilot studies on RI treatment systems for wastes from the paper industry in Montana with BOD_5 concentrations of 600 mg/L and applying hydraulic loadings of about 0.2 ft/day.

Some of the above-mentioned specialized arrangements had desirable outcomes with BOD_5 concentrations of 1000 mg/L or more. Thus, we can infer that neither BOD_5 nor COD can impose a limitation on designing urban land treatment systems. Table 6.6 summarizes the common organic loadings for frequently used applications.

Table 6.5 BOD_5 removal at typical land treatment systems

Process/location	Hydraulic loading, ft/year	BOD5, mg/L Applied	BOD5, mg/L Effluent	Sample depth, ft
Hanover, N.H. San Angelo, Tex	4–25 10	40–92 89	0.9–1.7 0.7	5
Rapid Infiltration				
Lake George, N.Y	140	38	1.2	10
Phoenix, Ariz	360	15	1.0	30
Hollister, Calif	50	220	8.0	25
Overland Flow				
Hanover, N.H	25	72	9	
Easley, S.C	27	200	23	
Davis, Calif	41	112	10	

Table 6.6 Typical organic loading rates for land treatment systems

Process	Organic Loading, lb BOD$_5$/(acre.day)[*]
Slow Rate (SR)	45–450
Rapid Infiltration (RI)	130–890
Overland Flow (OF)	35–100

[*]lb BOD$_5$/(acre.day) × 1.121 = kg/(ha.day)

6.2.4.3 Pathogenic Organisms

Some pathogens that can raise main concerns in land treatment systems include parasites, bacteria, and viruses. The main vectors of transmission that raise worries include reaching these pathogens to groundwater, polluting agricultural crops, farm animals feeding on contaminated plants, and spread of infections through airborne particles or via runoff. Certain measures including adsorption, dehydration, filtration, predation, and exposed to unfavorable conditions are among the common approaches adopted to eliminate pathogens in land treatment systems. Among the discussed processes, SR is the most effective way to eliminate five logs (10^5) of fecal coliforms from the soil of a few-feet depth. On the other hand, RI and OF treatment systems are able to eliminate 2 to 3 logs and 90% of fecal coliforms across an area of several feet.

6.2.4.4 Metals

Considering the elimination of metals from wastewater, the SR treatment system can yield the highest output, since the soil has a finer structure so that exposure and adsorption are more likely to be implemented. RI process can also be desirable in this sense; the difference is that it requires a longer distance of flowing overland due to the coarser structure of soil and higher hydraulic loading. Wastewater in OF systems has the least contact with soil, thus the removal capacity may vary from 60 to 90 percent depending on hydraulic loading and the considered metal. Clay minerals, metal oxides, and organic substances act as the basic surface on which adsorption of trace elements takes place. Therefore, the adsorption capacity in the case of fine and organic solid is relatively higher compared with that of sandy soils (Bagchi 1994; Delarestaghi et al. 2018).

The main issue that concerns most practitioners is the likelihood of these materials trapping within soil layers and subsequent diffusion through the food chain from crops and animals to human beings at the end of the cycle. The most critical metals are cadmium (Cd), lead (Pb), zinc (Zn), copper (Cu), and nickel (Ni). A specialized guideline on annual and cumulative additions of metals to agricultural lands has been developed by the World Health Organization (WHO) (Table 6.7).

Table 6.7 Metals concentrations in wastewaters and suggested concentrations in drinking and irrigation waters

Element	Raw sewage Mg/L	Drinking water Mg/L	Irrigation water, mg/L	
			20 years[a]	Continuous[b]
Cadmium	0.004–0.14	0.01	0.05	0.005
Chromium	0.02–0.70	0.05	20	5.0
Lead	0.05–1.27	0.05	20	5.0
Zinc	0.05–1.27	0.05	20	5.0

[a]For fine-textured soils only. Normal irrigation practice for 20 years
[b]For any soil, normal irrigation

6.2.4.5 Nitrogen

Considering the fact that various forms of nitrogen (N_2, organic N, NH_3, NH_4, NO_2, and NO3) exist and it easily undergoes the transition from one oxidation state to another, the elimination of nitrogen in land treatment systems appears as a complicated and dynamic task.

Specific forms of nitrogen (organic, ammonia, nitrate, etc.) and also total contents of them in the wastewater are required to be taken into account for the design of every process. The acquired experience from practical experiments indicates that the less oxidized nitrogen enters the land treatment system, the more efficient will be the operations of preservation and elimination of nitrogen (Pazoki et al. 2012).

The portion of ammonia is more likely to be lost by volatilization, taken up by the crops, or adsorbed by the clay minerals that are present in the soil. When conditions are desirable for nitrification (i.e. alkalinity and temperature are at suitable levels), it can be present within a range of 5 to 50 mg/(L.day) (Table 6.8).

If these reactions are supposed to have happened with ions of adsorbed ammonia within the top 4 inches of a given soil with fine structure, we can say that approximately 60 lb of ammonia nitrogen per acre will be converted to nitrate during each day. Nitrification is actually a nitrogen conversion process rather than a removal one. Removal is only possible through denitrification, volatilization, and crop uptake. In the case of SR systems, that is the crop uptake which accounts for the major way of removal. Nevertheless, the efficacy of denitrification and volatilization for this aim can be considerable when site conditions and wastewater types are suitable.

In the case of the RI process, ammonia adsorption on the soil particles usually takes place after nitrification; however, the most critical mechanism for removal is denitrification. Crop uptake, volatilization, and denitrification have the same contribution to nitrogen removal in OF systems.

Table 6.9 presents relevant data for mineralization rate with regard to wastewater biosolids. The given values refer to the percentage of the organic nitrogen that undergoes mineralization (conversion to inorganic forms such as ammonia and nitrate) during a year.

Table 6.8 WHO recommended annual and cumulative limits for metals applied to agricultural cropland

Metal rate[†]	Annual loading rate[a] lb/acre[‡]	Cumulative loading lb/acre[‡]
Arsenic	1.78	36.58
Cadmium	1.70	34.80
Chromium	133	2677
Copper	67	1338
Lead	13	268
Mercury	0.76	15.2
Molybdenum	0.80	16.1
Nickel	18.7	375
Selenium	4.5	89
Zinc	125	2498

[a]Loading lb/acre per 365-day period
[†]Cumulative loading over lifetime of site
[‡]lb/acre * 1.1208 = kg/ha

Table 6.9 Mineralization rates for organic matter in biosolids

| Time after biosolids application, years | Mineralization rate, % | | | |
	Unstabilized primary	Aerobically digested	Anaerobically digested	Composted
0–1	40	30	30	10
1–2	20	15	10	5
2–3	10	8	5	
3–4	5	4		

6.2.4.6 Phosphorus

Urban wastewater contains phosphorus in various forms including orthophosphate, polyphosphate, and organic phosphates. Orthophosphates can readily involve in biological reactions in soil ecosystems. The essential hydrolysis of the polyphosphates usually occurs with a slow rate in ordinary soils, so such forms are often absent (Qasim and Chiang 1994; Pazoki et al. 2015).

Plant uptake together with biological, chemical, and/or physical processes can take part in the task of phosphorus removal in land treatment systems. Phosphorus removal in soil considerably depends on nonrenewable chemical reactions. Therefore, the capacity for phosphorus preservation declines over time, not completely be lost.

RI systems lack crop uptake, so the effective phosphorus removal requires longer flowing distances considering specific properties of the soil and higher hydraulic loading rates.

In the case of OF systems, surface reactions determine the possibility of contact between the flowing wastewater and the soil, thus the rate of phosphorus removal may vary from 40 to 60 percent. Certain chemical additions and precipitation on the treatment slope can improve phosphorus removal in such systems.

6.2.4.7 Arsenic

Arsenic is a substance that is not necessary for any form of life. In high contents, arsenic is fairly toxic to plants and highly toxic to animals. Since toxicity caused by arsenic leads to unwanted effects in crops even at low levels, the food chain is a major concern in land treatment sites. Adsorption resulted from soil colloids with clay helps arsenic to be removed in the soil system, and iron and aluminum oxides play a similar role that they have in phosphorus removal.

Using poultry manure with contents of 15 to 20 ppm has been a common method of fertilizing for 20 years (0.2 to 0.4 lb As/(acre. year)), and no undesirable effect on alfalfa or clover has been reported so far. In the case of industrial sewages that have a significant content of arsenic, it is reasonable to conduct field examinations in order to define certain criteria in terms of loading rates and vegetation across the considered area.

6.2.4.8 Sulfur

Most wastewaters contain sulfur in the forms of sulfate or sulfite. Crop uptake has a critical impact on sulfur removal. Common values with regard to different crops are given in Table 6.10. It is reasonable to suppose that a major portion of sulfur compounds will form sulfate by mineralization. Taking drinking water aquifers into consideration, the allowable standard of 250 mg/L for sulfate in drinking water has to be defined within the project boundary. The harvested crop seems the major

Table 6.10 Sulfur uptake by selected crops

Crop	Harvested mass	Sulfur removed lb/acre
Corn	200 bu/acre	44
Wheat	83 bu/acre	22
Barley	100 bu/acre	25
Alfalfa	6 ton/acre	30
Clover	4 ton/acre	18
Coastal Bermuda grass	10 ton/acre	45
Orchard grass	7 ton/acre	50
Cotton	2.5 bale/acre	23

permanent removal procedure in sizing the system. Table 6.10 gives general values that can be used as general guidelines for making estimations.

6.2.5   Hydraulics of Soil Systems

Under certain conditions where the land is subjected to a hydraulic gradient, one of the major indicators in estimating the capability of the considered soil is the hydraulic conductivity. Hydraulic conductivity in the case of one-dimensional vertical flow is given by the following equation:

$$U = -K\frac{dh}{dz}$$

where U represents Darcy's velocity (that is the average velocity of the soil fluid through geometric cross-sectional area inside the soil), h represents the hydraulic head, and z represents the vertical distance within the soil. The coefficient of permeability and hydraulic conductivity are often used interchangeably. Hydraulic conductivity, based on Eq. 6.1, is defined as the ration of Darcy's velocity to the applied hydraulic gradient. Dimensions of K and velocity are equal in value which is length per unit of time (LT^{-1}).

There is a direct correlation between hydraulic conductivity and soil grain size, the structure of the soil matrix, the type of soil fluid, and the concentration of the fluid (saturation) within the soil matrix. The solid matrix of the soil has significant characteristics that include pore size distribution, pore shape, tortuosity, specific surface, and porosity. Certain properties including fluid density, ρ, and fluid viscosity, μ, are of high significance for the soil fluid. The hydraulic conductivity, K, in the case of a subsurface system saturated with the soil fluid is expressed by the following equation:

$$K = \frac{k\rho g}{\mu}$$

where, k, the intrinsic permeability of the soil, depends only on properties of the soil matrix, and g, the fluidity of the liquid, refers to the properties of the penetrating fluid. The hydraulic conductivity, K, the intrinsic permeability, k, and the fluidity, g, are expressed in terms of length per unit of time (LT^{-1}), l^2, and $L^{-1}T^{-1}$ respectively.

On the basis of Eq. 6.2, Darcy's law can be rewritten in a more explicit way in terms of its coefficient of proportionality (hydraulic conductivity K) as follows:

$$K = \frac{k\rho g}{\mu} = \frac{|U|}{|dh/dz|}$$

Table 6.11 Range of saturated hydraulic conductivity of various soil materials

Soil Type	Saturated Hydraulic Conductivity, K (m/yr)
Unconsolidated deposits	
Gravel	$1 \times 10^4 - 1 \times 10^7$
Clean sand	$1 \times 10^2 - 1 \times 10^5$
Silty sand	$1 \times 10^1 - 1 \times 10^4$
Silt, loess	$1 \times 10^{-2} - 1 \times 10^2$
Glacial till	$1 \times 10^{-5} - 1 \times 10^1$
Unweathered marine clay	$1 \times 10^{-5} - 1 \times 10^{-2}$
Rocks	
Shale	$1 \times 10^{-6} - 1 \times 10^{-2}$
Unfractured metamorphic and igneous rocks	$1 \times 10^{-7} - 1 \times 10^{-3}$
Sandstone	$1 \times 10^{-3} - 1 \times 10^1$
Limestone and dolomite	$1 \times 10^{-2} - 1 \times 10^1$
Fractured metamorphic and igneous rocks	$1 \times 10^{-1} - 1 \times 10^3$
Permeable basalt	$1 \times 10^1 - 1 \times 10^5$
Karst limestone	$1 \times 10^1 - 1 \times 10^5$

We can apply Eq. 6.3 to experimentally decide on the values of the intrinsic permeability, k, and the hydraulic conductivity, K, provided that fluid properties of density and viscosity are given. The amount of the saturated hydraulic conductivity in soils may widely vary over a range of magnitude in several orders that depend on the soil material. A general list of some anticipated values of K for different soil material including unconsolidated and consolidated is given in Table 6.11 below. Table 6.12 presents the anticipated values of K according to the grain size distribution, degree of sorting, and silt content of various soil materials in detail.

Table 6.12 Saturated hydraulic conductivities for fine-grained materials

Grain-Size Class	Saturated Hydraulic Conductivity K (10^3 m/yr)
Clay	< 0.0001
Silt, clayey	0.1–0.4
Silt, slightly sandy	0.5
Silt, moderately sandy	0.8–0.9
Silt, very sandy	1.0–1.2
Sandy silt	1.2
Silty sand	1.4

Values associated with saturated hydraulic conductivity can also represent variations across the space domain because the spatial forms are so versatile in the geological structure of soils.

6.2.5.1   Methods of Analysis

Field examinations and laboratory experiments can both be adopted in order to assess the saturated hydraulic conductivity of water in soil (or the intrinsic permeability of the soil). In any taken approach, measuring the K (or k) plays an important role in determining the numerical value of the mentioned coefficient in Darcy's equation.

Any adopted approach to determine K (or k) should follow this procedure:

1. Suppose a flow pattern (one-dimensional flow in a porous medium for instance) that can be explained analytically according to Darcy's law:

$$K = \frac{k\rho g}{\mu} = \frac{|U|}{|dh/dz|}$$

2. Implement an experiment replicating the chosen flow pattern and calculate all measurable values in Eq. 6.4 including fluid density, dynamic viscosity, flow velocity, and the gradient of the hydraulic head.
3. Substituting the quantified values into Eq. 6.4, calculate the coefficient K (or k).

The coefficient K (or k) can be obtained using different field or laboratory experiments.

American Society for Testing and Materials, the US Environmental Protection Agency, the US Department of the Army, and the US Department of the Interior have developed certain standard approaches to determine saturated hydraulic conductivity in soil materials: ASTM 1992a-o, EPA 1986, DOA 1970, and DOI 1990a,b respectively.

Generally, examinations that are conducted in laboratory situations involve small samples of the considered soil materials extracted from core-drilling operations. The resultant findings from these small samples are deemed as point representations of the soil characteristics. The calculated value of K (or k) can only be an accurate demonstration of the real saturated hydraulic conductivity at the considered point when the used samples are undisturbed.

Laboratory procedures can be used to assess the vertical and horizontal hydraulic conductivity in soil samples. For example, the values of K (or k) that are obtained from undisturbed samples of cohesive or cohesionless soils (frictional soil) correspond to sampling orientation that is generally vertical. In the case of disturbed (remolded) samples from cohesionless soils, the conductivity may be used to estimate the actual value of K in the undisturbed soil in the horizontal direction (DOA 1970). When the soil to be analyzed is of fine-grained type, the undisturbed cohesive sample can have a matching orientation in order to assess the hydraulic conductivity in either vertical or horizontal direction.

Field methods, compared with laboratory methods, usually require a relatively larger area of the soil. Therefore, the findings of the field methods represent impacts in both vertical and horizontal directions, and they indicate an average value of K. This approach is particularly significant for soils that are structured in different layers so that the values found for K may represent the most permeable layer in the soil profile. Nevertheless, if a specialized correct technique is adopted in the considered field, it will be possible to determine the in-situ values associated with vertical and horizontal components of K independently for each layer.

The objectives of the analysis will determine the specific method to be used in a particular application. Since obtaining undisturbed samples of unconsolidated soil may raise some difficulties, laboratory methods are usually unable to identify a K value that is an accurate demonstration of the considered value in the field. Thus, field examinations are more reasonable when the target of the investigation is to give an accurate delineation of the physical properties of the subsurface system. It is worth to note, however, that field methods are more costly than laboratory methods. Therefore, laboratory approaches may appear more preferable when considerations of cost are significant, actual demonstrations of the soil profile in the field are not so important, and there is no direct access to in-situ hydraulic conductivity.

6.2.5.2 Flow in the Horizontal and Vertical Directions

When we are to investigate a stratified soil with horizontal arrangement of the layers, the coefficient of permeability for each stratum may be different from the next depending on the direction of flow, and this is attributable to fabric anisotropy. Consider k_{h1}, k_{h2}, k_{h3}, and etc. as coefficients of permeability of layers 1, 2, 3, and etc. respectively, and also assume flow in the horizontal direction. Furthermore, let us assume k_{v1}, k_{v2}, k_{v3}, and etc. are coefficients of permeability for flow in the vertical direction. Thus, coefficients of total permeability of all layers in the horizontal direction can be obtained by the following equation:

$$K_{e(v)} = \frac{H}{\frac{H1}{k_{v1}} + \frac{H2}{k_{v2}} + \frac{H3}{k_{v3}} + \dots}$$

6.2.5.3 Laboratory Methods

In the laboratory settings, the examiner may use different instruments including permeameter, pressure chamber, and consolidometer (DOA 1970) to determine the value of K. All these methods share a common feature in the applied technique in which the soil sample is placed in a small cylindrical receptacle reflecting a one-dimensional structure through which a fluid is forced to flow.

(a) Constant-Head Method

The most frequently used method in laboratory settings to determine the saturated hydraulic conductivity of soils with coarse grains is the constant-head test using a permeameter. This test directly complies with Darcy's law for soil liquid configuration that demonstrates a one-dimensional and consistent flow of a liquid that penetrates into a saturated column of soil from an unvarying cross-sectional area. This approach is carried out on a cylindrical sample of soil with cross-sectional area A and length L. The prepared sample is placed between two porous plates that lack excess hydraulic resistance to flow. H_2-H_1 is the constant head difference which is applied across the test sample. The saturated hydraulic conductivity K of the soil is directly calculated by Darcy's equation as below once the volume V of the test fluid flowing through the system during the time t has been estimated:

$$K = \frac{VL}{At(H_2 - H_1)}$$

It is reasonably recommended that the test be iterated several times taking varying head differences into account in order to obtain more accurate results. Moreover, the quantity of the collected liquid should be much enough to make feasible taking at least three significant figures from the measured volume. Constant-head permeameter in a simpler arrangement involves a lower boundary of K measurement which is about 1×10^1 m/yr; it is almost equivalent to lower boundaries of conductivity for soils of sandy clay type. When we are to analyze relatively lower values of K, it is better to apply either an intensified version of the constant-head permeameter (i.e. a more sensitive version to volume flow rate measurements) or a permeameter of falling-head type.

(b) Falling-Head Method

This is a specific test using a permeameter in laboratory settings to determine K (or k) value for fine-grained soils. Falling-head method has common features with the constant-head one with regard to applying Darcy's law directly to the same one-dimensional and saturated column of soil with a consistent cross-sectional area. In this method, a cylindrical soil sample with cross-sectional area A and length L is placed between two plates with high conductivity. The soil sample column is attached to a standpipe with the cross-sectional area a through which the disseminating fluid enters the system. Therefore, by calculating the head variations from H_1 to H_2 in the standpipe during a predefined interval of time t, the saturated hydraulic conductivity is obtained by the following equation:

$$K = \left(\frac{aL}{At}\right) Ln\left(\frac{H_1}{H_2}\right)$$

The lower boundary of K, estimated by a falling-head permeameter is approximately 1×10^{-2} m/yr. This quantity is almost equivalent to the lower boundary of conductivity for silts and –coarse-grained clays.

The main drawback that is the case in both approaches is associated with the degree of saturation obtained within the soil samples. The empty space in the porous area usually contains trapped air bubbles that although disappear gradually by dissolving into the deaerated water, the pressure exerted by them can deviate the obtained results. Consequently, special care is needed in using the test instruments aiming at K measurements. The degree of the sample saturation has to be validated by estimating the volumetric content of water in the sample, then to compare the obtained results against total porosity estimated based on the particle density.

To have more accurate estimations of K in laboratory settings where the trapped air bubbles may cause more critical conditions, it is recommended to measure the conductivity with backpressure. This method involves extra pressure (back pressure) that is applied to the pore fluid of the soil sample. Taking this measure can minimize the size of air bubbles within the pores so that the degree of water saturation increases to an acceptable level.

6.2.5.4 Field Methods

We can categorize the field methods arranged for determination of saturated hydraulic conductivity of soils into two main classes: (1) methods that are devised for sites adjacent to or below a shallow water table and (2) methods that are applied to sites either above a deep water table or where there is not any water table. In other words, these methods are aimed at the sites located in saturated and unsaturated areas of the soil in question. Regardless of the approach adopted to these purposes, Darcy's law acts as the main reference based on which the value of K is determined once the gradient of the hydraulic head at the site and the resulting soil water flux have been estimated.

(a) Field Methods Applied on Saturated Zones of the Soil

A wide range of field methods has been devised to determine the saturated hydraulic conductivity of saturated soils within groundwater structure taking both confined and unconfined conditions. The methods are of two types: (1) the auger-hole and piezometer methods that are applied under the conditions of the unconfined shallow water table and (2) well-pumping tests which are essentially directed towards a determination of aquifer properties associated with the development of confined and unconfined groundwater systems.

(b) Field Methods Applied on Unsaturated Zones of the Soil

The task of applying field methods for estimating the saturated hydraulic conductivity of unsaturated soils that are above the water table (or when a water table is absent) is relatively difficult compared with measuring K for saturated soils. What differentiates them lies in the fact that in order to make estimations, the primary unsaturated soil requires to be saturated via an artificial way. The quantity of water needed to saturate the medium is considerably large, so the task of measurement will be more challenging and time-consuming. The findings of such field estimations of K represent field-saturated hydraulic conductivity.

There is a wide range of field methods that have been specially invented for the determination of the field-saturated hydraulic conductivity of soil materials within the unsaturated (vadose) zone of the soil. Among the existing methods designed for making estimations of field-saturated K, certain methods are regarded as standard approaches: (1) the shallow-well pump-in or dry auger-hole, (2) the double-tube, (3) the ring infiltrometer, (4) the amount of air, and (5) the constant-head test in a single drill hole. You can find the detailed guidelines for comparing these standard methods in ASTM D5126-90 under the title of Standard Guide for Comparison of Field Methods for Determining Hydraulic Conductivity in the Vadose Zone (Amoozegar et al. 1986).

6.2.6 Identifying and Choosing the Site

In some situations where choosing an appropriate process correlates with selecting a site for treatment, a two-phased planning procedure is the most common. Figure 6.1 depicts these two phases. Phase one entails primary estimation of land requirements with regard to properties of wastewater and climate characteristics, identifying potential sites within the considered region, investigating the sites according to technical and economic factors, and finally choosing the potential sites. Having the potential sites been chosen, phase two entails field evaluations, primary estimations on design and cost, contrasting among available alternatives, and finally choosing the most cost-effective option.

6.2.6.1 Storage Requirements

Certain conditions such as cold weather, precipitation, or crop management necessitate sewage storage. To make initial estimations, climatic factors (e.g. freezing days of the year) will be enough to be taken into account in order to verify the storage requirements. In the case, wastewater treatment systems of SR type designed for agricultural crops, the crop management time with regard to harvesting and planting operations should be added to the storage days. As far as RI and forested SR systems are under consideration, at least 7 to 14 storage days may be preferred for making initial estimations about the region.

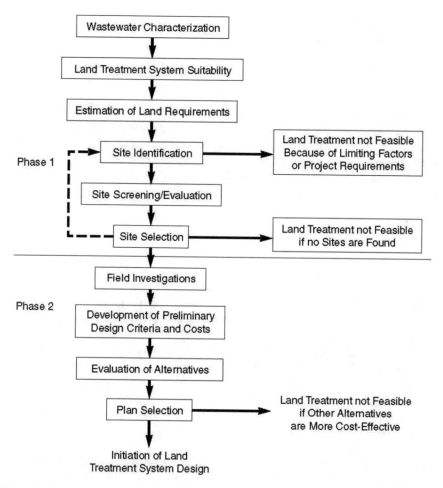

Fig. 6.1 Two-phase planning process

6.2.6.2 Site Area Estimations

The primary requirements associated with the site area can be estimated regarding wastewater flows, storage needs, and initial loading rates. The equation below demonstrates the relationship between field area, loading rates, and operating period.

$$F = 13,443 \frac{Q}{LP}$$

where,
    F: field area, in acres.
    Q: average flow, in mgd.

L: loading rate, in inch/week.
P: the period of application, in weeks/year

$$13443 = conversionfactor = 3.069\frac{acre * ft}{mgd} * \frac{12\ in * 365\ days}{year}$$

In order to estimate the period of application, we should subtract the storage period from 52 weeks/year. Table 6.4 summarizes site areas for a 1 mgd flow in relation to all three systems. The values associated with SR and RI systems include a 20% extra area in addition to the estimated field area to take unusable land into account as well. This additional amount is 40% in the case of OF systems for accounting extra inefficiency in building overland flow slopes.

6.2.6.3   Topography

Sharp slopes may impose some restrictions on potentials of a site where there is a significant rate of runoff and erosion, crop cultivation has more difficulties, and also saturation in such areas may bring about unstable conditions for the soil in question. The allowable gradient is determined with regard to the properties of the soil and the treatment process that is to be applied.

The next topographical issue that deserves attention is the relief that refers to a height difference between two points across a land treatment region. The most critical effect of relief on the treatment system deals with the cost of conveying wastewater to the desired site. In most situations, making use of pumping systems to convey wastewater to a nearby site have to be compared to building gravity waterways to a faraway site in terms of cost-effectiveness (Table 6.13).

The recurrence rate of flooding within an area is another important aspect of choosing a site. The major factors that are to be considered in estimating the risk of flooding are the potential severity of a flood, frequency of flood occurrence, and areal extent of flooding.

6.2.6.4   Soils

Drainage is not so satisfactory in fine-grained soils that hold water for a longer period of time. Therefore, compared with loamy soils with good draining properties, percolation will be slower and crop management will be more difficult. The most desirable option for the purposes of applying OF process in the land treatment system is a soil that has a fine texture (Tables 6.14 and 6.15).

The most favorable option for the SR process is loamy or medium-grained soil. Nevertheless, certain crops can grow well in quick-drain soils. Sandy soils are appropriate for the RI process. Table 6.16 present favorable ranges of permeability and suitable textures separately for each process.

Table 6.13   Site identification land requirements

System	Land requirements acres/mgd	
Slow rate—agricultural		
No storage	200	
1 month's storage	225	
2 month's storage	250	
3 month's storage	275	
4 month's storage	315	
5 month's storage	350	
6 month's storage	415	
Slow rate—forest		
No storage	310	
1 month's storage	335	
Rapid infiltration		
Primary effluent	30	
Secondary effluent	15	
Overland flow		
Storage (months)	Screened wastewater	Secondary effluent
0	90	180
1	100	200
2	110	220
3	120	240
4	140	280

Table 6.14   Land use suitability factors for identifying land treatment sites

Land use factor	Type of system			
	Agricultural slow rate	Forest slow rate	Overland flow	Rapid infiltration
Open or cropland	High	Moderate	High	High
Partially forested	Moderate	Moderately high	Moderate	Moderate
Heavily forested	Low	High	Low	Low
Built upon (residential, commercial or industrial)	Low	Very low	Very low	Very low

Table 6.15 Grade suitability factors for identifying land treatment sites

Grade factor %	Slow rate systems		Overland flow	Rapid infiltration
	Agricultural	Forest		
0–12	High	High	High	High
12–20	Low	High	Moderate	Low
20	Very low	Moderate	Eliminate	Eliminate

Table 6.16 Typical soil permeabilities and textural classes for land treatment processes

	Land treatment processes		
	Slow rate	Rapid infiltration	Overland flow
Soil permeability range, in/h	0.06–2.0	> 2.0	< 0.2
Permeability class range	Moderately slow to moderately rapid	Rapid	Slow
Textural class and range	Clay loams to sandy loams	Sandy and sandy loams	Clays and clay loams
Unified soil	GM-d, SM-d, ML, OL, MH, PT	GW, GP, SW, SP	GM-u, GC, OL, CH, OH, SM-u, SC, CL

6.2.6.5 Geology

Preliminary estimations need to take geological formation into account. Inconsistencies or cracks over the bedrock may lead to short-circuiting of the flow or can disturb the predictable flow patterns.

6.2.6.6 Site Selection

Site selection and evaluation stage can only begin when the relevant data about the site properties have been collected and mapped. When we have a few sites with relatively known suitability, the best site can be chosen by a simple analysis of cost-effectiveness. However, when we are to deal with myriad options, a site screening process will be necessary.

6.2.7 Types of Natural Treatment Systems

What follows is a more detailed description of the three main natural treatment systems that were introduced earlier.

6.2.7.1 Slow Rate Process

This process is a controlled way of applying wastewater over land covered by vegetation. The rate in this system is usually measured in terms of a few inches for one week (see Fig. 6.2). The waterway is typically designed according to infiltration, percolation, and later flow within the range of treatment. Treatment is accomplished on the surface while the sewage percolated through the matrix of the vegetation roots and soil. The most portion of the wastewater is used by the plants, a certain part of it reaches the groundwater, and the remaining may be recovered for other purposes. Any typical design tries to avoid off-site runoff the wastewater.

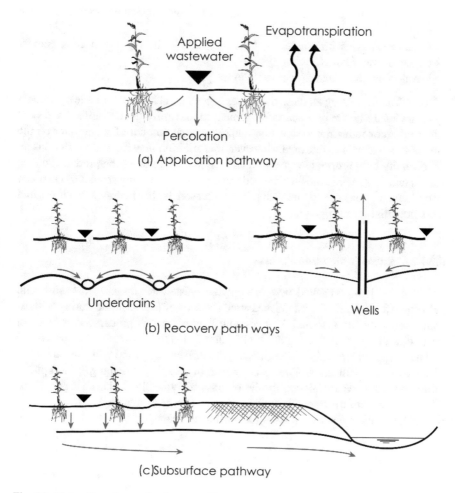

Fig. 6.2 Hydraulic pathways for slow rate (SR) land treatment

Possible designs of hydraulic pathways for wastewater can include:

– Irrigation of plants with a growing rate of percolation aimed at salt leaching,
– Vegetative uptake percolating through the main path,
– Directed percolation towards drains or wells for purposes of water recovery,
– Directed percolation towards groundwater and/or lateral subsurface stream to nearby surface waters.

Possible ways of applying wastewater may include irrigation through furrows or border strip flood or using fixed and movable sprinkler systems. The existing conditions in the considered site determine the application mode. The vegetation on the land surface plays a significant role in all SR systems. The main objectives of the SR process are:

– Sewage treatment
– Gaining profits from water recovery and taking advantage of nutrients to produce crops
– Replace sewage with usable water for irrigation in dry regions meeting general targets of water conservation
– Developing green spaces accessible to the public

The ideal is realizing all these objectives together, but it is not possible by a single system. Generally, the optimum arrangement to maximize the cost-effectiveness of either urban or industrial systems is to apply a large amount of wastewater on the land of as small as possible area. However, this criterion may impose restrictions on the diversity of appropriate vegetation and the market value of the produced crops. Humid climates require more optimized treatment. Long-lasting species of grass can be optimum options due to longer application season, higher hydraulic loadings, and more potential for nitrogen removal.

6.2.7.2   Rapid Infiltration Process

This process is a controlled way of applying wastewater on earthen basins with permeable soils. Here, the rate is measured in terms of feet of liquid during a week. Compared with SR systems, the hydraulic loading in this process is an order of magnitude higher. Considering this high hydraulic loading, vegetation on the surface will have a peripheral effect. Nevertheless, the surface vegetation will play an important role in the stabilization of the exterior layer of the soil and keeping the acceptable infiltration rates. Water-tolerant species of grass are typically used in such situations. RI process enables treatment through biological, chemical, and physical interactions in which the top layers of soil from the most active part.

The waterway in this process is designed in a way that enables surface infiltration, subsurface percolation, and lateral sideways stream (see Fig. 6.2). The application cycle entails a flooding period and days or weeks of drying period afterward. This cycle provides the infiltration surface with enough time to be restored aerobically, and also allows the percolation to be drained. Among the mentioned processes, RI sites require more careful considerations in terms of geo-hydrological aspects, so the accurate design depends on a proper definition of subsurface conditions and the local groundwater structure.

RI system is mainly aimed at treating wastewater. Although all arrangements and operations are directed toward the realization of this objective, there are some alternatives for the final disposal or reusing the runoff following treatment:

- Groundwater recharge
- Recovery of the treated water for subsequent reuse or discharge
- Recharge of nearby surface streams
- Seasonal storage of treated water within subsurface layers of the site to be recovered for seasonal agricultural purposes.

Recovery and reuse of the post-treatment runoff have particular importance in dry climates. Several studies in Arizona, California, and Israel concluded that recovery of the treated water has beneficial effects for irrigation. Groundwater recharge is also an interesting issue, but great caution is needed for nitrogen management if aquifers of potable water are involved. In the case of applying urban sewage, it is almost impossible that potable water levels for nitrate-nitrogen (10 mg/L as N) can be achieved immediately under the treatment zone unless the appropriate measure is taken. Otherwise, the wastewater needs to be sufficiently mixed and dispersed with the native groundwater before it arrives at the end of the collection slopes. When the operation is to be accomplished in humid areas, there is no need for recovery or reuse considerations.

In such situations, it only suffices that the RI site is located near a body of surface water to avoid undesirable groundwater effects. The quality of the subsurface stream entering the surface water is often much better than that of an advanced treatment plant outflow.

6.2.7.3 Overland Flow Process

Overland flow (OF) is another controlled way of applying wastewater on lands with relatively penetrable soils and over grasslands with a gentle gradient. Here, the hydraulic loading is usually several inches for one year, and it is much higher compared with SR systems.

Considering pertinent costs directly dependent on hydraulic loading, OF systems are usually more cost-effective than SR systems assuming the same requirements for water quality.

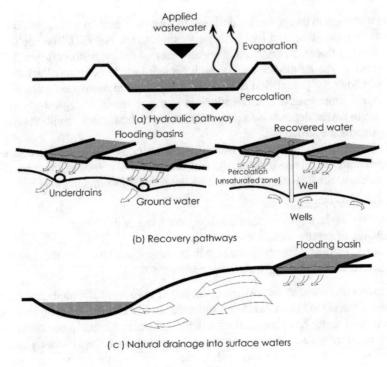

Fig. 6.3 Hydraulic pathway for rapid infiltration (RI)

Long-lasting species of grass can play a critical role in the OF systems because this kind of vegetation can both stabilize the soil over sloping lands and prevent erosion.

The waterway in this process is designed in such a way that allows sheet flow down the specially arranged vegetated surface, and runoff is collected in ditches or drains at the bottom of the slopes (see Fig. 6.3). The treatment process is accomplished while the applied wastewater interacts with the soil, the plants, and the biological growths on the surface. Most treatment reactions resemble those that occur in trickling filters and other dependent growth processes. In this process, gated pipes or nozzles at the top of the slope or sprinklers installed over the sloping surface are utilized to apply the wastewater. Sprinkler systems are more common for industrial wastewaters and sewages with a high content of solids (Qasim and Chiang 1994; Christensen et al. 2005; Pazoki et al. 2015).

Some portion of wastewater may be lost due to deep percolation and evapotranspiration, but a large portion of it will be collected at the bottom of the slopes and will then discharged into nearby surface water. Surface discharge is almost an inevitable component of OF systems which require pertinent permissions. The main purpose of the overland flow process is to treat wastewater in a cost-effective way. The harvest and sale of the surface crops may bring additional benefits that will compensate for the costs. One of the largest sites in the US designed for 5 mgd flow of urban sewage treatment has been established in Davis, California (Fig. 6.4).

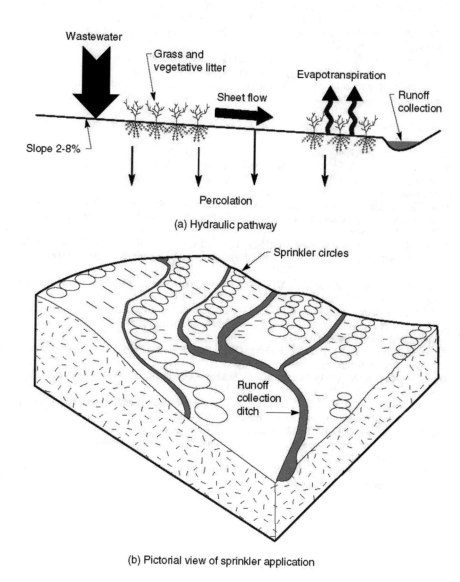

(a) Hydraulic pathway

(b) Pictorial view of sprinkler application

Fig. 6.4 Hydraulic pathways for overland flow (OF)

References

Amoozegar A, Warrick AW, Fuller WH (1986) Movement of selected organic liquids into dry soils. Hazardous Waste Hazardous Mater 3(1):29–41

Bagchi A (1994) Design, construction, and monitoring of landfills

Bhargava AKSHEY, Lakmini SHAMA (2016) Land treatment as viable solution for waste water treatment and disposal in India. J Earth Sci Clim Change 7:375

Christensen TH, Cossu R, Stegmann R (Eds.) (2005) Landfilling of waste: leachate. CRC Press

Crites RW, Middlebrooks EJ, Bastian RK (2014) Natural wastewater treatment systems. CRC press

Delarestaghi RM, Ghasemzadeh R, Mirani M, Yaghoubzadeh P (2018) The comparison between different waste management methods of Tabas city with life cycle assessment assessment. J Env Sci Studies 3(3):782–793

Ghasemzade R, Pazoki M (2017) Estimation and modeling of gas emissions in municipal landfill (Case study: Landfill of Jiroft City). Pollution 3(4):689–700

Ghasemzadeh R, Pazoki M, Hoveidi H, Heydari R (2017) Effect of temperature on hydrothermal gasification of paper mill waste, case study: the paper mill in North of Iran. J Env Studies 43(1):59–71

Mulamoottil G, McBean EA, Rovers F (2018) Constructed wetlands for the treatment of landfill leachates. Routledge

Oweis IS, Khera RP (1990) Geotechnology of waste management

Pazoki M, Abdoli MA, Ghasemzade R, Dalaei P, Ahmadi Pari M (2016) Comparative evaluation of poly urethane and poly vinyl chloride in lining concrete sewer pipes for preventing biological corrosion. Int J Env Res 10(2):305–312

Pazoki M, Abdoli M, Karbasi A, Mehrdadi N, Yaghmaeian K, Salajegheh P (2012) Removal of nitrogen and phosphorous from municipal landfill leachate through land treatment. World Appl Sci J 20(4):512–519

Pazoki M, Delarestaghi RM, Rezvanian MR, Ghasemzade R, Dalaei P (2015) Gas production potential in the landfill of Tehran by landfill methane outreach program. Jundishapur J Health Sci 7(4)

Pazoki M, Ghasemzade R, Ziaee P (2017) Simulation of municipal landfill leachate movement in soil by HYDRUS-1D model. Adv Env Technol 3(3):177–184

Pazoki M, Ghasemzadeh R, Yavari M, Abdoli M (2018) Analysis of photocatalyst degradation of erythromycin with titanium dioxide nanoparticle modified by silver. Nashrieh Shimi va Mohandesi Shimi Iran 37(1):63–72

Pazoki M, Pari MA, Dalaei P, Ghasemzadeh R (2015) Environmental impact assessment of a water transfer project. Jundishapur J Health Sci 7(3)

Qasim SR, Chiang W (1994) Sanitary landfill leachate: generation, control and treatment. CRC Press

Renou S, Givaudan JG, Poulain S, Dirassouyan F, Moulin P (2008) Landfill leachate treatment: review and opportunity. J Hazardous Mater 150(3):468–493

Shayesteh AA, Koohshekan O, Khadivpour F, Kian M, Ghasemzadeh R, Pazoki M (2020) Industrial waste management using the rapid impact assessment matrix method for an industrial park. Global J Env Sci Manag 6(2):261–274

Tchobanoglous G, Theisen H, Vigil S (1993) Integrated solid waste management: Engineering principles and management Issues. McGraw-Hill

Chapter 7
Solid Waste Management in Tehran

7.1 Introduction

Municipal waste management formally started in 1911 by the establishment of the Tehran municipality. At that time people disposed of their solid wastes in simple ways. The aim of their act was just to transport the waste out of the urban area. This situation continued until the last two decades. At the begging of the 1980s authorities concentrated more on municipal waste management in the big cities and the recycling centers were began to be developed as the result (Abdoli and Azimi 2010).

Before 1980, wagons and tricycles were the most regular waste collection method used by the sweepers. They usually collected the organic wastes and sold them to the farmers as the fertilizer (Abdoli and Azimi 2010; Abduli et al. 2011). In the big cities, people submitted their wastes to the big wagons or tricycles which were carried by mules and horses. On the other hand, in smaller cities, the generated wastes usually were transported directly by the people to the outside of the cities (Abdoli and Azimi 2010).

In the next years, the amounts of generated waste were increased rapidly by the population growth and consequently, the common collection methods such as wagons and tricycles were not efficient anymore. Thus, by developing the industries and entering vehicles to Iran, the waste collection systems were improved. Municipalities widely used open trucks to collect and transport the waste. The most prevalent disposal method in most of the cities was open dumping (Abduli 1996; Pazoki et al. 2018).

In the 1970s, there was an enormous immigration wave from the rural area to the cities. Consequently, the percentage of urban population was increased eminently and the municipal waste generation was quickly raised (Abdoli 2010). At the end of the 1970's, mechanized waste management systems started to developed and used in big cities specifically Tehran. Most of the changes were in the collection methods and using the vehicles. Since then, many advances in various waste management have occurred (Abdoli and Azimi 2010).

© Springer Nature Switzerland AG 2020
M. Pazoki and R. Ghasemzadeh, *Municipal Landfill Leachate Management*,
Environmental Science and Engineering,
https://doi.org/10.1007/978-3-030-50212-6_7

In this chapter, the related research and projects about waste management in Tehran and Mashhad are gathered and discussed.

7.2   Municipal Waste Management in Iran

7.2.1   Tehran

Tehran, the capital city of Iran with a population of 8.2 million. The municipality of Tehran is responsible for the solid waste management of the city; the waste is mainly landfilled in three centers in Tehran, with a small part of it usually recycled or processed as compost. Nevertheless, an informal sector is also active in collecting recyclable matters from solid waste. The municipality has lately initiated some activities to mechanize solid waste management and reduce waste production (Pazoki et al. 2015). The waste produced in Tehran can be divided into five main groups (Rupani et al. 2019):

– Residential waste (residential, administrative, commercial, educational, etc.)
– Outdoor waste (cleanup of the streets and parks by the districts municipal authorities themselves) (Fig. 7.1)
– Industrial waste
– Health-care waste
– Agricultural waste

Fig. 7.1 Location of Tehran in Iran

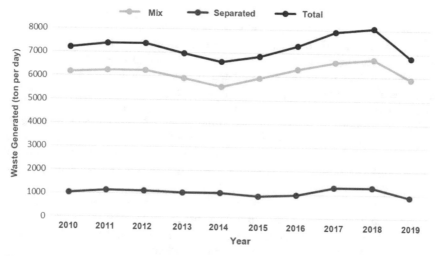

Fig. 7.2 Daily avarage of mixed and dry wastes and their total duing 2010–2019

Most of the municipal solid waste in Tehran is mixed (combined dry and wet waste). As can be seen from Fig. 7.2, the amount of mixed waste plus dry waste follows a relatively similar trend over the 2010–2019 period (Abdoli and Azimi 2010).

Statistics indicate that the amount of waste is produced by every Tehran citizen annually reaches to 6 times their weight. The waste produced per day in the world is between 250 and 300 g per capita while it is 600 g in Iran and 1200 g in the north of Tehran. In general, the central districts and district 20, where the majority of business centers are located, generate the highest amount of waste. Statistics show that business districts produce more waste and more valuable waste. The production of solid waste in urban districts between the years 2001 to 2005 had a 1.1% increase. The quantity of waste produced in the north of Tehran, based on the studies conducted, is at least twice as much as the rest of the state and 4 times as much as the world standard. In Tehran, waste is collected two or three times a day, whereas in other states waste is collected two or three times a week (Abdoli and Azimi 2010).

The spatial analysis of waste production in Tehran shows that the greatest amount of waste is produced along the north–south axis. This axis is the backbone of the historical Tehran, where most the businesses are located. Thus, the greatest amount of waste is produced not by families but by transactions. In urban communities, the quantity of waste produced is related to the lifestyle of the people and their type of activities. The analysis of the lifestyle of the families living in each district explains the nature and amount of the waste they produce. Also, the amount of waste produced is related to the consumption pattern of packed food and drinks. As the consumption of such food and beverages is higher in northern than western districts, it is obvious that more waste is produced in these districts. Since more buildings in northern districts have central heating systems, in comparison with southern districts, the people living in northern districts will find it easier to present fuel and typically consume more fuel. Therefore, more waste is generated in the northern districts.

The spatial analysis of valuable waste produced in 2007 shows that the higher amount of valuable waste was produced along north–south and east–west axes, where business activities are mainly concentrated. It should be noted that valuable waste is the result of commercial packing and unpacking, which is responsible for the highest quantity of commercial waste. Collection, transmit and segregation of waste in Tehran are done by Recycling Institution, which has established a network of recycling centers in all parts of the city. The waste is collected at smaller centers located in urban districts and is transferred to bigger centers. The spatial division of these centers shows that smaller ones are located in residential areas and bigger ones at non-residential spaces outside the city (Figs. 7.3 and 7.4).

There important challenges in managing solid waste in Tehran which include: collection and management of industrial and health-care waste; public education to reduce and separating household waste and educating municipal workers in order to optimize the waste collection system; and the participation of the private sector in solid waste management.

Due to the below Figure the amount of solid waste generated in Tehran between 1990 and 2009 is given. As can be viewed it has an upward trend. The recent amount of municipal waste generation in Tehran is about 8000 tons per day which is the highest waste production rate in Iran.

One of big problems in managing of solid waste is to handle a large amount of waste (7.641 tons/day in 2008) in Tehran. In 2010, Abdoli and Azimi studied the diversified possible strategies of waste management methods in Tehran. As their studies, source reduction can be introduced as one of the first priorities for solid

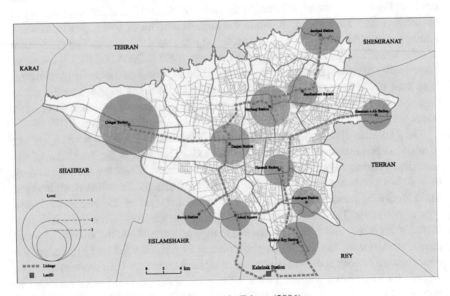

Fig. 7.3 Transfer stations of waste management in Tehran (2006)

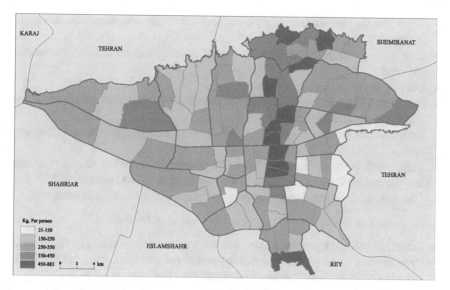

Fig. 7.4 Map of Tehran based on per capita waste production during the year (2006)

waste management in Tehran (Abdoli and Azimi 2010). This study indicates that the first attempt to quantify the source reduction potential in the city, and subsequently, outlines the principle guidelines, legislations, and strategies regarding source reduction application in Tehran metropolitan area. Based on the findings source reduction policies can be implemented in dealing with packaging material, paper, street waste, mixed household waste, and hazardous household wastes (Abdoli and Azimi 2010; Parsa et al. 2018).

Also, other wastes like industrial wastes produced inside city boundaries can be reduced by implementing source reduction measures. It is also found that any recycling program can be mixed effectively with source cut strategies (Zand and Abduli 2008; Shayesteh et al. 2020). The waste reduction potential for each component of waste stream is calculated as a result of the research source reduction potential were determined a horticultural waste, 80%; food waste, 80; paper and cardboard, 50%; textiles, 20%; metals including ferrous and nonferrous, %90; Glass, %30; PET%, 70, and plastic 80%. The overall potential for source reduction in Tehran city is estimated to be 66% for the waste stream as a whole (Abdoli and Azimi 2010; Delarestaghi et al. 2018).

Abdoli & Azimi in 2010 suggest some key and strategic issues in suggested Tehran source reduction laws which are noted as follows (Abdoli and Azimi 2010):

– To determine an organization accountable for planning, supervision, and coordination of source reduction and determine the organization's authority
– Explaining the waste generation in each section and deciding on the regarding source reduction potential
– Conjecturing the costs due to the cost-saving goals

- Reporting the performance of different sections to aim to source reduction to the responsible organization
- Cooperating with different industries to modify their products and set the strategies required
- Levying taxes on disposal products
- Providing tax credits or exemptions to industries that meet set source reduction goals in design and produce
- Set the source reduction database in each section (household, industrial, commercial)
- Implementing scientific researches and technologies regarding source reduction and preventing the negative influence on environmental and economic
- Controlling, supporting and modifying program with international strategies

Also, the key elements of the proposed law are as following:

- Incentivize efforts to reduce, re-use, recycle waste and recover energy from waste;
- Reform rules to drive the reduction of waste and diversion from landfill while reducing costs to compliant businesses and the regulator;
- Target action on materials, products, and sectors with the greatest scope for improving environmental and economic results;
- Stimulate investment in collection, recycling and recovery infrastructure, and markets for recovered materials that will maximize the value of materials and energy recovered; and
- Elevating the national, regional and local governance, with a clearer performance and institutional framework to deliver better-coordinated action and services on the ground.

In Fig. 7.5 the recommended organizational chart of waste reduction in Tehran is illustrated (Fig. 7.6).

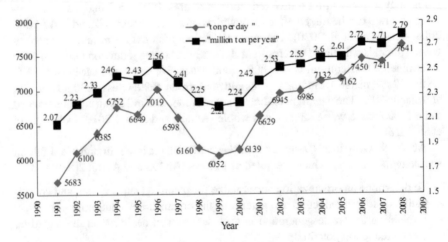

Fig. 7.5 Total solid waste generation in Tehran from 1991 to 2008

Fig. 7.6 The recommended organizational chart of waste reduction in Tehran

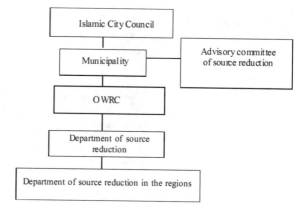

Aradkouh sites us the main landfill site in Tehran. Almost, most of Tehran's solid waste is transported to this landfill and handled. In 2010, a series of sampling and physical analysis are conducted to evaluate the Municipal Solid Waste (MSW) components and characteristics in Aradkouh center (Alimohammadi et al. 2010). The results are presented in Table 7.1.

As can be viewed, the amount of organic and wet solid waste is high which the proper methods of leachate management systems need to be considered.

In 1996, the production of industrial wastes in the Tehran metropolitan area is increasing in quantity and variety as a result of technological developments and rapid industrialization (Abduli et al. 2007). The quality of the Tehran environment is a matter of growing concern to the city authorities and the importance of efficient solid waste management is increasingly recognized. As a result, in 1988, the Budgetary and Planning Organization and the Department of Environment called for studies on industrial waste management systems (Figs. 7.7, 7.8, 7.9 and 7.10).

One of the main problems in this baseline study was the lack of adequate and accurate data. The need for a centralized and integrated system of waste data recording with emphasis on the quantity and quality of waste generated in different sectors should be highlighted. The data available in many sectors, especially in the dry waste sector, is highly imprecise and inconsistent. Thus, proper planning for at source separation and recycling is very difficult with current available data. Currently, only mixed waste inputs to the Araad Kouh Landfill Site are documented and are relatively accurate. Other information such as outdoor waste, dry waste, hospital waste, and industrial waste are relatively inaccurate.

7.2.2 Mashhad

Tehran is not the only megacity in Iran that has acute problems in SWM. Mashhad is another big city in Iran that every year faces a huge amount of tourists in a limited

Table 7.1 MSW components and characteristics in Aradkouh center

Waste types	Weight %
Wet waste	67.8
Bread	1
Soft plastic	2.2
Hard plastic	0.6
PET	0.7
Plastic bags	6.2
Paper	4.4
Cardboard	3.7
Ferrous metals	1.6
Non-ferrous metals	0.2
Textile	3.4
Glass	2.4
Wood	1.7
Tires	0.7
Leather	0.6
Dust and rubble	1.3
Special waste (health care waste)	1.6
Density	450 kg/m^3
Moisture content	40–44%
Waste temperature	35 °C
pH	6.0–7.4

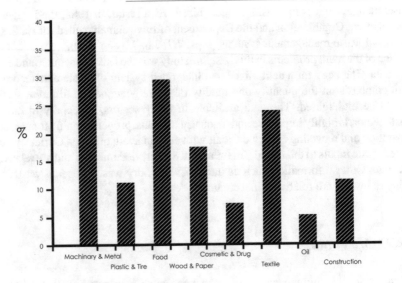

Fig. 7.7 Type of industrial products of Tehran

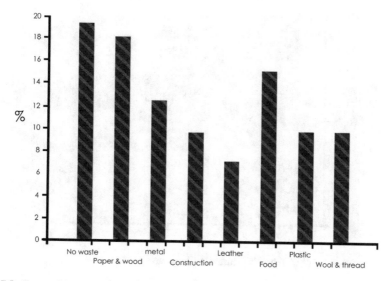

Fig. 7.8 Compositions of industrial wastes in Tehran's industries

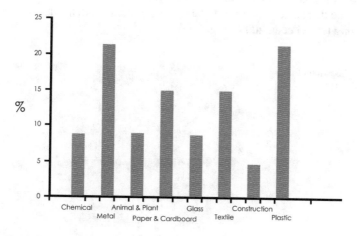

Fig. 7.9 Type of recovered materials in Tehran's industries

period of time. The total amount of solid waste generated in Mashhad in 2008 was 594, 800 tons with per capita solid waste generation rate of 0.609 kg person^{-1} day^{-1}. The amount of recycled dry solid waste was augmented from 2.42% of total dry solid waste (2588.36 ton.year^{-1}) in 1999 to 7.22% (10, 165 tons.year^{-1}) in 2008. The most important fractions of recycled dry solid waste in Mashhad included paper and board (51.33%), stale bread (14.59%), glass (9.73%), ferrous metals (9.73%), plastic (9.73%), polyethylene terephthalate (2.62%) and non-ferrous metals (0.97%).

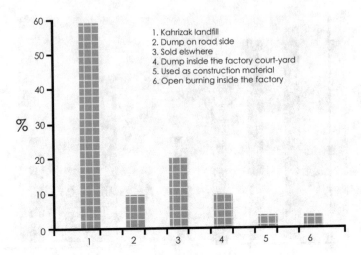

Fig. 7.10 Disposal sites of Tehran's industrial solid waste

Hopelessly the potential of dry solid waste recycling in Mashhad has not been viewed properly and there is a great effort to be made in order to earn the desired situations of recycling (Noori et al. 2009) (Fig. 7.11).

Fig. 7.11 Location of Mashhad in Iran

7.3 Environmental Costs

In 2010, Abdoli et al. worked on a theoretical framework to pin down the environmental costs, benefits, and the net welfare effects associated with hazardous waste management. Their research reviewed and presented a theoretical model to determine the costs, merits, and welfare effects of hazardous wastes management (Abduli 1996; Abdoli et al. 2010).

Based on Iranian law, environmental costs are assigned to waste producing firms. However, in practice, due to weak enforcement programs, companies do not pay any environmental costs. Using the basic principles and logic of welfare economics, we present a micro-level model for analyzing an industry that produces waste as a by-product of its production process. Firms in the industry choose the least cost method of disposal (either legal or illegal disposal). By thinking through various figures of presented models in partial equilibrium mechanism we found $R' R'$ which are the net welfare effect of producing firms, the net welfare effect to firms supplying legal waste disposal services and, R'_1 and R'_2 and 3the net welfare effect of the environmental damage, respectively. By analyzing the presented figures we concluded that state regulatory policy may ideally lower environmental costs through a subsidy program (Bozorg-Haddad et al. 2020; Pazoki et al. 2017; Ghasemzade and Pazoki 2017).

7.4 Composting

There are lots of studies on composting systems and a solid waste management program and the environmental impacts of this method in Iran.

Tajbakhsh et al. (2008), investigated the recycling of spent mushroom compost using earthworms Eisenia feotida and Eisenia Andrei. A 90-day study conducted to explore the potential of epigeic earthworms Eisenia foetida and Eisenia andrei to transfigure the different types of agricultural wastes and spent mushroom compost into a value-added product, i.e., vermin compost. Vermicomposting ended in a significant reduction in C: N ratio, pH, electrical conductivity, total organic carbon, TK; and increase in total Kajeldahl nitrogen, TP, and several micro and macronutrients compared to those in the worm feed. Our trials proved that the vermin composting could be considered as an alternative technology for recycling and environmentally safe disposal/management of the mushrooms cultivation complexes' residues merged with different types of agricultural waste using epigeic earthworms E. foetida and E. Andrei (Tajbakhsh et al. 2008).

Vermicomposting systems can potentially help to convert the wastes into value-added materials, avoiding their widespread disposal which has environmental and economic drawbacks; this process will reduce the costs related to the exceptional use of different types of agricultural wastes as feeds for earthworms. The information presented here should provide a sound basis for the management of SMC, mixed with other agricultural wastes, in large commercial vermin-composting systems.

The vermin-composts obtained in this study were rich in total nitrogen, phosphorus, and other essential factors for plants' growth, and had good physical properties, low conductivity, low C:N ratios, optimal stability, and maturity. These features make vermicomposts useful as soil conditioners, healthy organic fertilizers, and good substitutes in potting media (Tajbakhsh et al. 2008).

Also in 2008, Rezaei et al. studied the bioremediation of TNT contaminated soil by composting with municipal solid wastes, Soil and Sediment Contamination. In recent years, a number of biological processes have been utilized for bioremediation of TNT contaminated soils. The goal of this research was to select the proper formulation of the compost materials based on mechanically sorted organic fraction of municipal solid wastes (MSW) that can provide the proper balance of carbon and nitrogen and the best conditions to promote biological activity in the compost and to reach the highest TNT removal. Seven factors, including TNT, contaminated soil (%w/w), trace element solution (%v/w), chicken manure (%w/w), TNT concentration (mg/Kg), microbial suspension (%v/w) of isolated and selected microorganisms from TNT contaminated sites, cow manure (%w/w), and composting condition (aerobic and anaerobic) were considered using Taguchi statistical methods. L8 and L9 orthogonal arrays and analysis of variance (ANOVA) were applied to assess the most significant factors affecting the TNT removal. The optimum conditions to maximize TNT removal were aerobic condition, 35% (w/w) TNT contaminated soil, 5% (w/w) cow manure, 5% (v/w) microbial suspension and 50,000 (mg/Kg) TNT concentrations. In optimum conditions, TNT removal of 99.99% was obtained within 15 days, which is a relatively short time when compared with previous studies. The very high organic contents (over 70%) of municipal solid wastes in Iran provide ample raw materials for the production of compost to treat contaminated soils (Rezaei et al. 2010).

7.5   Leachate Management in Iran

Unfortunately, the use of an engineering and sanitary waste disposal centers in Iran is not focused sufficiently. Not using this center will cause extensive economic and environmental issues. For instance, a twelve-acre lake with a depth of eighty feet can be seen in the Kahrizak landfill is a giant consequence of these issues (Abduli and Safari 2003; Rashidi et al. 2012). Thus, studying the research in the developing states and modify them for the local condition to obtain a new management system can be a proper solution. In addition, it should be noted that most of the cities in Iran have dry or semi-dry weather with little raining days. Therefore, the concentration of organic matter in the leachate is high which is different from the conditions of western countries (Fig. 7.12).

In the majority of countries, sanitary landfill sites are viewed to be the most frugal and commonly used disposal method for municipal solid wastes, and also industrial waste. However, landfills suffer from downsides such as the creation of

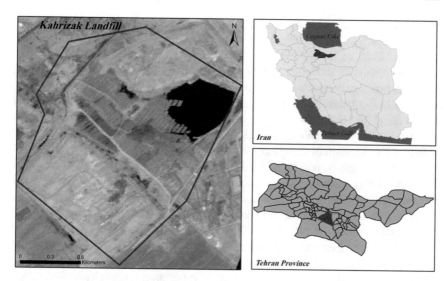

Fig. 7.12 Aradkouh (Kahrizak) landfill

utterly contaminated leachates, eminent changes in the volumetric flow and chemical composition of leachate.

The leachate contains dissolved and suspended organic and inorganic pollutants and toxic chemicals. Leachate can contaminate groundwater, surface water, and soil. In addition, it can have detrimental health effects on human beings. It is customary that leachate filtration methods in Iran are adopted based on the proposed ways in western countries. There is a dire need to exploit a different view infiltration and treatment of leachate in Iran based on the climatic differences, available land, the quality of leachate, and technological and financial resources (Abduli and Safari 2003; Pazoki et al. 2016) (Table 7.2).

The moisture content of municipal solid waste in Iran is about 60–70% and is higher than typical values in Europe and the United States. Therefore, it is predicted that the main part of leachate in Iran's landfill will be generated by excessive water

Table 7.2 Raw leachate's characteristics in Aradkouh's landfill

Parameter	Standard Method	Range	Average ± S.D
COD (mg^{-1})	5220–COD	40,000–70,000	$50,000 \pm 2400$
BOD_5 (mg^{-1})	$5210–BOD_5$	20,000–30,000	$27,000 \pm 1700$
TDS (mg^{-1})	2540–TDS	16,500–18,000	$17,000 \pm 500$
TSS (mg^{-1})	2540–TSS	20,300–26,200	$22,000 \pm 4000$
TOC (mg^{-1})	2310–TOC	16,500–20,000	$18,000 \pm 1600$
EC (ms/cm)	2510–EC	27.3–33.3	30.6 ± 3.3
pH	$4500–H^+$	4.8–5.1	5 ± 0.3

content. Hydrologic Evaluation of Landfill Performance (HELP) model is one of the most approved instruments to simulate the hydrological attributes of landfills. Although, some major deviations from real values have been reported about the calculated results for leachate generation by the HELP model other researchers and/or engineers practically have used it in some places to estimate the quantity of leachate produced in the landfills. On the other hand, this model is elaborated and mainly used in developed countries with the waste having low moisture content and also in climatic conditions with high precipitation (Alimohammadi et al. 2010).

In 2010, Alimohammadi et al. studied the applicability of the model in arid areas, by the construction of two 30 m * 50 m (effective horizontal length) test cells in Kahrizak landfill, and monitoring the real leachate generation from each one. A set of field capacity and saturated water conductivity tests were also performed to determine the basic hydrologic properties of municipal waste landfilled. A comparison was made between values calculated by HELP model and recorded quantities, shows that a prediction of leachate on annual basis can be done by HELP model with acceptable accuracy but when the infiltration of water to waste body goes up as a result of leachate production, the model intends to underestimate water storage capacity of the landfill, which lead to deviation of calculated quantities from real ones (Alimohammadi et al. 2010) (Fig. 7.13).

Direct evaporation from waste is one of the widespread phenomena for landfill leachate production and compost management in arid and semi-arid regions. Due to this, in another study Abdoli et al. investigated the direct evaporation from the waste body and its influence on leachate production in landfills in arid areas.

In this study, a newly established test was performed to determine a correlation between pan evaporation and waste surface evaporation. A container with a diameter of 120 cm and a height of 60 cm is used for determining the evaporation in field conditions in Kahrizak Landfill in Tehran based on the container weight loss. A class A evaporation pan is also installed in the vicinity to determine evaporation from the water surface. A correlation could be seen in the results and also a certain diminishing

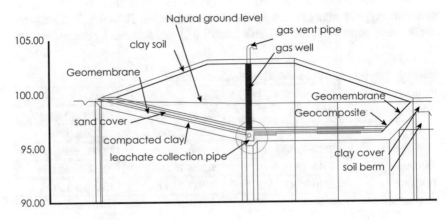

Fig. 7.13 The schematic presentation of test cells (section)

by time pattern in evaporation from the waste is also observed. A governing equation is also provided to calculate direct evaporation from waste using pan evaporation values. The correlation shows with acceptable precision with a 0.5 mm day^{-1} error. A comparison also made with the previous researches in Germany shows higher values for maximum evaporation per day (8 mm day^{-1} instead of 5 mm day^{-1}) and higher maximum possible values for evaporation (180 mm instead of 20 mm). The Current research shows that evaporation can have a significant impact on the amount of leachate produced in landfills in arid areas and implementing the current tool will lead to the more efficient design of leachate treatment in landfills or better decision making for operation of landfills (Alimohammadi et al. 2010).

According to the abovementioned if we consider a landfill with average pan evaporation of 12 mm, only after 65 days the DEW will be, less than 0.1 mm. What they can deduce from that is in the arid areas, the evaporation will not stop before 2 months if no rewetting has taken place which is quite different from areas with a wet climate. This was also indicated the importance of evaporation in landfill operation. For example, if we consider a landfill with 1 Ha operative front (if a thin layer operation is considered that is quite prevalent for a medium-size landfill) and 12 mm pan evaporation if for 3 days no landfill cover is applied, the amount of 226 m^3 potential reduction in the volume of the leachate will take place on daily basis. This may lead to a considerable reduction in the size of leachate treatment facilities. A comparison to Braunschweig—Germany's experience shows also original new findings. First, the rate of daily evaporation can surpass 5 mm/day in arid areas up to 8 mm/day as reported in the DEW tests (Alimohammadi et al. 2010; Pazoki et al. 2014). On the other hand, accumulative evaporation will surpass the quantity of 20 mm which was reported before in Germany.

Current Dew tests show the capability of evaporation more than 180 mm. One should also have this fact in mind that as the methodology for evaluation of the evaporation is different in these two types of research that may be responsible for some parts of difference. In general, the dimension of the test cells in Germany was bigger less evaporation is expected from them. Based on current research a tool to assess the direct evaporation of the waste body is developed that shows the considerable difference to dominant trends in developing countries. This may be because of the high moisture content of waste which contributes to evaporation during the time and also stronger sunlight radiation in arid areas. Current analysis can be completed by additional tests regarding different phases of evaporation and the effect of very high evaporation rates (for example in the south of Iran). At the present level of studies, different phases on evaporation can be observed although further research needs to be done regarding the boundaries of different phases and their relationship with pan evaporation value. The current equation in hand can is determined based on the specific condition of the Karizak Landfill site in the south of Tehran but it is believed that it can be applied in any other area with highly wet municipal waste and positive evaporation budget.

The leachate includes dissolved and suspended organic and inorganic, toxic or chemical pollutants. Leachate can contaminate groundwater, surface water, and soil. It can additionally have harmful health effects on a human being.

The treatment of municipal landfill's leachate is considered as one of the most significant environmental issues. Leachate may be treated by evaporation to attain acceptable discharge limits for various types of contaminants. One of the most important operational problems with evaporation is scaling. Therefore, in a study which was conducted by Afsharnia et al. in 2012, the sono-evaporation treatment method was studied (Afsharnia et al. 2012).

As ultrasound can affect many leachate compounds, this study was done to test its influence on scale treatment using evaporation. In order to determine the effect of sonication on leachate evaporation, some leachate samples were sonicated at a frequency of 25 kHz and 500 W. Sonication periods were tested in the range of 10–40 min on leachate samples. Evaporation and distillation were performed on leachate samples. The tests demonstrated that 90% of the leachate was distilled. When evaporation–distillation processes were carried out, chemical oxygen demand (COD) reduction was 83.56%. However, when sonication was used, COD reduction was 86.56–88.36% for 10–40 mi sonication, respectively. During evaporation, a part of total Kjeldahl nitrogen (TKN) remained in the concentrate and the rest entered into the distillate. If sonication was applied, TKN reduction was 83.70 and 89.71% for 10 and 40 min, respectively. Sonication had no outstanding effect on EC and heavy metals reduction. When using sono-evaporation, the efficiency of scale removal (ESR) reached 95.97% (Afsharnia et al. 2012).

No data have been published on the leachate quality for the Kahrizak landfill site, including heavy metal concentrations. Thus, concentrations of Cu, Zn, Pb, Cd, Ni, Co, Fe, and Mn were measured in four fresh and older leachate samples during a 10 months period. Conceptually, heavy metals can be considered to be present in three phases: (a) Dissolved; (b) Adsorbed onto suspended solids, and (c) adsorbed onto the suspended solids with particle sizes less than 0.45 m.

In research by Abdoli and Safari (2003) a preliminary analysis of heavy metals in the Kahrizak landfill leachate was conducted. The method proposed in that measured the concentrations of heavy metals in each of the three steps (Abduli and Safari 2003) (Table 7.3).

Fresh leachate with a lower pH value potentially undergoes biochemical reactions; thus, the pH is increased. Heavy metals as a part of leachate constituents show higher solubility in fresh leachate than in mature and/or older leachate. Suspended solids mainly originate from the daily cover soil wash out into the leachate mixture. The soil type of suspended solids is the central parameter defining the status of heavy metals in a particular step. The higher the clay content, the higher the consolidation of heavy metals in a particulate step. Thus, in Abdoli and Safari's study, primary heavy metal precipitation and suspended solids omission can potentially decrease the concentration of suspended solids and heavy metals, both in the particulate and dissolved steps. As the pH of the leachate increases, the concentration of particulate heavy metals increases too. Thus it is proposed to adjust the pH around 6.5 to 7.5 artificially (e.g. by the addition of hydrated lime), followed by sedimentation of the suspended solids, before any treatment process. In the former procedure, heavy metals can be precipitated and at the same time, an appropriate pH value for biological reactions can be achieved, while the latter process decreases the potential of heavy

Table 7.3 Concentration of heavy metals in three steps of **a** dissolved, **b** particulate and **c** colloidal in samples No. 1 and 2 (ppm)

Heavy metals	Sample No. 1			Sample No. 2		
	Dissolved	Particulate	Colloidal	Dissolved	Particulate	Colloidal
Cu	0.107	10.38	0.024	0.271	8.565	0.204
Zn	0.211	56.396	1.240	1.546	119.239	1.318
Pb	ND	ND	0.05	0.206	ND	0.588
Cd	0.03	1.748	0.033	0.134	2.207	0.073
Ni	0.283	13.989	0.345	1.017	8.985	0.322
Co	0.111	8.654	0.164	0.481	4.898	0.374
Fe	8.128	1181.894	6.965	12.035	960.070	1.782
Mn	0.062	103.625	4.518	7.039	109.612	0.252

metal adsorption in case pH is changed as a result of anaerobic processes (e.g. increasing pH in the acid formation phase of anaerobic degradation) (Abduli and Safari 2003).

Another leachate treatment procedure is natural treatment systems. The natural treatment systems can be categorized into two systems, that is, soil systems and water systems. The soil systems or land treatment systems are the treatment systems that use the leachate on the land surface, under controlled conditions, to earn a definite level of filtration and treatment through physical, chemical and bioprocesses in the matrix of water-soil–plant. The land treatment method requires a lower level of energy in comparison with other widespread filtration and treatment methods. In this filtration system, merely the moving and spreading the leachate requires energy.

Land treatment is considered to be a process of engineering in which contaminants are removed by the help of plants and soil. This procedure may lead to greater benefits thanks to the possibility it creates to supply agriculture and aquaculture with the necessary nutrients. Hence, its advantages outweigh the advantages of wastewater recycling. In the majority of land treatment operations, the wastewater undergoes primary and secondary treatment before application. These treatments include mechanical screening, settling, aeration, and biological decomposition. There are three delivery methods when we want to apply the wastewater to land including overland flow, rapid filtration, and slow-rate irrigation. Through a well-designed system, elements in the applied wastewater can be absorbed by the soil, transformed by soil micro-organisms, or be taken up by the plants' roots (Pazoki et al. 2014).

Among the above-mentioned methods, a slow irrigation system is the best method to run the process of land treatment on leachates. This system enjoys a high potential for removal and is able to omit a substantial amount of BOD, solid wastes, and deposition. The pre-treatment for land treatment and filtration method is to remove greater waste in the form of filtration and sedimentation. In addition, these prevent any harm to irrigation facilities and machinery (Pazoki et al. 2014).

The processes recruited in the land treatment of leachate include physical processes such as sedimentation and filtration, chemical procedures such as surface absorption, ion exchange, oxidation, deposition, as well as biological procedures such as biodegradation, microbial turnover, and plant absorption.

The release of nutrients such as P and N into the environment will be transported in surface waters, nitrate leachates into groundwater, ammonia emissions, and greenhouse gases (NO_X) can be seriously detrimental because of their capability of leaving unfavorable effects on the quality of waterways, biodiversity, public health, and the climate.

Plants' absorbing and uptake the nutrients can be a significant way to omit nutrients from the environment. Of course, it requires periodic harvesting and removal of biomass. In some cases, 650 and 100 kg/ha of N and P have been done by annual and woody species of plants.

The technology of using plants in the purification, refinement, preservation, and sustaining soil, sediments, and contaminated waters is both well accepted and used in the world. Phytoremediation is the general term for various methods via which plants are used to remove water and soil contamination. Moreover, herbs can degrade bio-contaminants. They can also act as a filter and a trap to absorb and stabilize heavy metals. One of the plant species which can be successfully utilized to do the above-mentioned procedures is the permanent grass called Vetiver, or Chrysopogonzizanioides. It originates from the Indian peninsula and first was used by the World Bank to preserve soil and water in India thanks to its remarkable features.

Pazoki et al. in 2014 investigated the attenuation of Municipal landfill Leachate through Land treatment. In their study, a laboratory experiment was conducted through land treatment, achieving an efficient and economical method by using the Vetiver plant. Moreover, the effects of land treatment of leachate of municipal landfills on the natural reduction of organic and inorganic contaminants in the leachate after the pre-treatment in the Aradkouh disposal center are invested. Three pilots including the under-investigation region's soil planted by the Vetiver plant, the region's intact soil pilot and the artificial composition of the region's soil including the natural region's soil, sand, and rock stone are used. The leachate, having passed its initial treatment, passed through the soil and to the pilot. It was collected at the end of the pilots and its organic and inorganic contaminants were measured. However, the land treatment of leachate was carried out at a slow rate at various speeds. According to the results, in order to remove COD, BOD5, TDS, TSS, TOC the best conclusion was obtained in the region's soil planted with Vetiver plant and at the speed of 0.2 ml per minute which resulted 99.1, 99.7, 52.4, 98.8, 94.9% removal efficiencies, respectively. It also can be concluded that the higher the organic rate load is, the lower the efficiency of the removal would be. Moreover, EC & pH were measured and the best result was obtained in the region's soil planted with the Vetiver plant and at the speed of 0.2 ml/min (Pazoki et al. 2014).

Likewise, they focused on the removal of Nitrogen and Phosphorous from municipal landfill leachate through land treatment. This research aims to investigate the effects of land treatment of municipal landfill's leachate on the natural reduction of nitrogen and phosphorous density existing in the leachate after the pre-treatment in the Aradkouh disposal center.

Three pilot studies were utilized to run the present investigation. The pilots included the under-investigation region's soil planted by the Vetiver plant pilot, the region's intact soil pilot, and the artificial composition of the region's soil including the natural region's soil, sand, and rock stone. The leachate, having passed its initial treatment, passed through the soil and to the pilot. It was collected at the end of the pilots and its nitrogen and phosphorus were measured. Nevertheless, the land treatment of leachate was conducted in a slow speed range at speeds of 0.2, 0.6, and 1 ml per minute through three repeated times. According to the results, in order to remove phosphate, the best result was obtained in the region's soil planted with the Vetiver plant and at the speed of 0.2 ml per minute which resulted in 98.77% of phosphate removal. Moreover, in order to remove nitrogen (organic nitrogen, ammonia, and nitrate), the best possible result was obtained in the region's soil planted with the Vetiver plant and at the speed of 0.2. Moreover, a substantial 93.44% reduction in ammonia and the increase of nitrate was reached (Pazoki et al. 2014) (Fig. 7.14).

As previously mentioned, the purpose of this investigation was to study the capability of land filtration and treatment in a natural reduction of different forms of nitrogen and phosphate existing in the leachate after the pre-treatment stage. The results were illustrated in the previous section. Drawing a comparison between these results will highlight the followings:

1. The greater the organic loading rate is, the lower the efficiency of the removal would be. In the lower hydraulic load, the rate of nitrogen and phosphorous is higher because the hydraulic retention time is longer.

Fig. 7.14 Stages of moving sewage from storage to pilots

2. By means of the Vetiver plant in leachate, land treatment increases the efficiency of the treatment. It also plays a very substantial role in the omission of nitrogen and phosphorous at an appropriate rate (To omit nitrogen and phosphorous, advanced treatment is needed which increases the expenses).
3. In this type of filtration, only the transmission of leachate from the output of the pre-treatment system, and spreading and spraying it on the surface requires some energy. It also reckons fewer mechanical facilities. Also, it needs lower and easier levels of maintenance compared to advanced treatment.

Producing energy is another method to handle the leachate. A significant amount of biogas can be produced in the anaerobic digestion step in the treatment plant reactor. Based on the high quantities of organic material in Iran's generated leachate, generating energy is efficient and can be considered significantly. 7500 tons of municipal solid waste which are generated in Tehran city is landfilled in the Kahrizak disposal site. The leachate of the waste has created a lake with 180000 m in volume.

Due to Rashidi et al. (2012), a power plant working with the leachate was modeled. The leachate organic load (BOD = 34,400 mg/L and COD = 53,900 mg/L) is considerably higher when compared with other countries due to the higher amount of organics available in Tehran wastes. The results indicate that an amount of 33,504 m^3/d biogas can be produced in Tehran's landfill that eventually would be sufficient to run a power plant of 3.4 MW capacities. The plant which is designed by Thermo flow software consists of two gas turbine units with 2 MW capacities so that the total capacity is 4 MW. About 10% of the generated power is for in-plant consumption and the rest can be sold. The results of the air pollution modeling using Screen 3 software indicate that CO and PM amounts are in allowed range but N2O exceeds the standard limits. The high temperature of the outlet gases emitted from gas turbines makes it possible to warm up water and regulate the temperature of the anaerobic reactor (Rashidi et al. 2012).

The biogas amassed in Kahrizak leachate treatment contains 58% methane according to the results obtained from the calculations. Moreover, methane volume produced in the anaerobic digester reactor is 37 m per one cubic meter of leachate. Since the inlet leachate rate is 1400 m^3/d, the produced biogas and methane rates in the anaerobic units will be respectively 33,504 and 19,433 m^3/d. 2.4 kWh electrical powers are generated in the plant for each cubic meter of biogas. So, a plant with 4 MW capacity can be constructed which is composed of two gas turbine units each having 2 MW capacities. Methane gas is consumed in the turbines and CO is produced which decreases the greenhouse effect of the gases. The produced power in the plant is 29,784*10^2 kWh annually.

Thermal energy due to outlet gases can be used to control the temperature in anaerobic digester reactors and also for the treatment plant staff consumption. Exhaust polluting gas emission modeling reveals that the less the temperature of the outlet gases is, the more the maximum amounts of the pollutants are and also the farther the pollutants critical point are located. In the meantime, the comparison between modeling results and clean climate standards indicates that emission amounts of CO and PM are in the range of clean climate conditions in Iran and comply with the standards of US environmental protection agency (EPA) however N2O pollutants rate is more than allowed limit. Some managerial and technical approaches like decreasing the height of the stack can be used to resolve the problem.

To pursue Rashidi's study, Abdoli et al. investigated the electricity from the Leachate treatment plant. In this research, the feasibility of electricity generation using biogas has been investigated. Considering that 68.81% of the waste is degradable, the produced leachate has a high organic load (COD = 53,900 mg/L and BOD = 34400 mg/L) (Abdoli et al. 2012). The results indicated that a power plant with a capacity of 1.8 MW could be constructed on the site. This electricity can be utilized in Kahrizak Disposal Site and also sold to the network (10 US cents/kilowatt). Financial analysis by means of ProForm software indicates 1.3 years of payback period and emission reduction of carbon dioxide equal to 5752 tones/year in comparison with the natural gas power plant. Thus, this project is financially feasible for private investors with an internal rate of return equal to 77% or more (Abdoli et al. 2012; Pazoki et al. 2015).

Physical analysis of Tehran MSW is shown in the below table. The existence of 68.81% of the biodegradable materials in the Tehran solid waste mixture shows the high potential of leachate generation in landfill sites. In addition, since the solid waste has high organic and biodegradable matters, the produced leachate has high organic loads. So based on the ingredients, water moisture rate was measured 50% and the waste chemical formula was figured out: $C_{555.8}H_{856}O_{277.3}N_{15.9}S$ (Tables 7.4, 7.5 and 7.6).

Kahrizak Leachate Treatment Plant is the largest one in the Middle East. The potential of biogas production at kahrizak leachate is $18m^3$ methane/m^3 leachate. Calculations indicated that biogas and methane production in an anaerobic unit of Kahrizak leachate treatment plant with $1400m^3$ flow rate was $29897m^3$ and $19433m^3$

Table 7.4 Physical analysis of Tehran municipal solid waste

Components	Mass%
Food waste	68.81
Papers	4.41
Cardboards	3.72
Rubbers	0.71
Plastics	8.9
Pets	0.71
Textiles	4.04
Debris	2.07
Woods	1.66
Glasses	2.4
Metals	2.56

Table 7.5 Energy results

	Annual average GJ $\times 10^3$	Total project GJ $\times 10^3$
Baseline fuel savings-nautral gas	146	2186
Project fuel inputs-biogas	128	1913

Table 7.6 Avoided emissions results

Pollutant	Annual average (tones/year)	Total project (tones/15 years)
Carbon dioxide	5752	86,281
NO_x	12	176

per day, respectively. Thus, constructing a power plant with a capacity equal to 1.8 MW can generate clean electricity, and the payback investment will be about 1.3 years with an internal rate of return equal to 77% or more (Abdoli et al.2012; Ghasemzadeh et al. 2017).

References

Abdoli MA (2010) A theoretical framework for determining environmental costs, benefits, and the net welfare effects associated with hazardous waste management. Caspian J Env Sci 8(2):195–202

Abdoli MA, Azimi E (2010) Municipal waste reduction potential and related strategies in Tehran

Abdoli M, Ramke H, Ghiasinejad H (2010) Direct evaporation from the waste body and its influence on leachate generation in landfills in arid areas. J Biolog Sci 10(2):107–111

Abduli MA (1996) Industrial waste management in Tehran. Env Int 22(3):335–341

Abdoli MA, Karbassi AR, Samiee ZR, Rashidi Z, Gitipour S, Pazoki M (2012) Electricity generation from leachate treatment plant

Abduli MA, Nabi BGR, Nasrabadi T, Hoveydi H (2007) Evaluating the reduction of Hazardous waste contact in Tabriz petrochemical complex, focusing on personal protective equipment method

Abduli MA, Naghib A, Yonesi M, Akbari A (2011) Life cycle assessment (LCA) of solid waste management strategies in Tehran: landfill and composting plus landfill. Env Monit Assess 178(1–4):487–498

Abduli MA, Safari E (2003) Preliminary analysis of heavy metals in the Kahrizak landfill leachate: a conceptual approach. Int J Env Stud 60(5):491–499

Afsharnia M, Torabian A, Mousavi GR, Abduli MA (2012) Landfill Leachate treatment by sono-evaporation. Desalinat Water Treat 48(1–3):344–348

Alimohammadi P, Shariatmadari N, Abdoli MA, Ghiasinejad H, Mansouri A (2010) Analysis of help model application in Smi-Arid areas, study on Tehran test cells. Int J Civil Eng 8(2):174–186

Bozorg-Haddad O, Mohammad-Azari S, Hamedi F, Pazoki M, Loáiciga HA (2020, June) Application of a new hybrid non-linear Muskingum model to flood routing. In Proceedings of the institution of civil engineers-water management (vol. 173, No. 3, pp. 109–120) Thomas Telford Ltd

Delarestaghi RM, Ghasemzadeh R, Mirani M, Yaghoubzadeh P (2018) The comparison between different waste management methods of Tabas city with life cycle assessment assessment. Journal of Environmental Sciences Studies 3(3), 782-793

Ghasemzade R, Pazoki M (2017) Estimation and modeling of gas emissions in municipal landfill (Case study: Landfill of Jiroft City) Pollution 3(4):689–700

Ghasemzadeh R, Pazoki M, Hoveidi H, Heydari R (2017) Effect of temperature on hydrothermal gasification of paper mill waste, case study: the paper mill in North of Iran. J Env Stud 43(1):59–71

Noori R, Abdoli MA, Ghasrodashti AA, Jalili Ghazizade M (2009) Prediction of municipal solid waste generation with combination of support vector machine and principal component analysis: a case study of Mashhad. Env Prog Sustain Energy: An Official Public Am Instit Chem Eng 28(2):249–258

Pazoki M, Abdoli MA, Ghasemzade R, Dalaei P, Ahmadi Pari M (2016) Comparative evaluation of poly urethane and poly vinyl chloride in lining concrete sewer pipes for preventing biological corrosion. Int J Env Res 10(2):305–312

Pazoki M, Abdoli MA, Karbassi A, Mehrdadi N, Yaghmaeian K (2014) Attenuation of municipal landfill leachate through land treatment. J Env Health Sci Eng 12(1):12

Pazoki M, Delarestaghi RM, Rezvanian MR, Ghasemzade R, Dalaei P (2015) Gas production potential in the landfill of Tehran by landfill methane outreach program. Jundishapur J Health Sci 7(4)

Pazoki M, Ghasemzade R, Ziaee P (2017) Simulation of municipal landfill leachate movement in soil by HYDRUS-1D model. Adv Env Technol 3(3):177–184

Pazoki M, Ghasemzadeh R, Yavari M, Abdoli M (2018) Analysis of photocatalyst degradation of erythromycin with titanium dioxide nanoparticle modified by silver. Nashrieh Shimi va Mohandesi Shimi Iran 37(1):63–72

Parsa M, Jalilzadeh H, Pazoki M, Ghasemzadeh R, Abduli M (2018) Hydrothermal liquefaction of Gracilaria gracilis and Cladophora glomerata macro-algae for biocrude production. Biores Technol 250:26–34

Pazoki M, Pari MA, Dalaei P, Ghasemzadeh R (2015) Environmental impact assessment of a water transfer project. Jundishapur J Health Sci 7(3)

Rashidi Z, Karbassi AR, Ataei A, Ifaei P, Samiee-Zafarghandi R, Mohammadizadeh MJ (2012) Power plant design using gas produced by waste leachate treatment plant. Int J Env Res 6(4):875–882

Rezaei MR, Abdoli MA, Karbassi A, Baghvand A, Khalilzadeh R (2010) Bioremediation of TNT contaminated soil by composting with municipal solid wastes. Soil and Sedi Cont 19(4):504–514

Rupani PF Delarestaghi R M, Abbaspour M, Rupani MM, EL-Mesery HS, Shao W (2019) Current status and future perspectives of solid waste management in Iran: a critical overview of Iranian metropolitan cities. Env Sci Poll Res, 1–13

Shayesteh AA, Koohshekan O, Khadivpour F, Kian M, Ghasemzadeh R, Pazoki M (2020) Industrial waste management using the rapid impact assessment matrix method for an industrial park. Global J Env Sci Manag 6(2):261–274

Tajbakhsh J, Abdoli MA, Goltapeh EM, Alahdadi I, Malakouti MJ (2008) Recycling of spent mushroom compost using earthworms Eisenia foetida and Eisenia andrei. The Environ 28(4):476–482

Zand AD, Abduli MA (2008) Current situation of used household batteries in Iran and appropriate management policies. Waste Manag 28(11):2085–2090

Printed in the United States
by Baker & Taylor Publisher Services